The Multiple Realization Book

The Multiple Realization Book

Thomas W. Polger &
Lawrence A. Shapiro

OXFORD
UNIVERSITY PRESS

Great Clarendon Street, Oxford, OX2 6DP,
United Kingdom

Oxford University Press is a department of the University of Oxford.
It furthers the University's objective of excellence in research, scholarship,
and education by publishing worldwide. Oxford is a registered trade mark of
Oxford University Press in the UK and in certain other countries

Published in the United States of America by Oxford University Press
198 Madison Avenue, New York, NY 10016, United States of America

British Library Cataloguing in Publication Data
Data available

ISBN 978-0-19-877989-6

To our children,
multiple but each uniquely realized,
even the triplets

Contents

Preface

It is often said that René Descartes set the agenda for philosophy of mind and psychology in the seventeenth century. Classes and textbooks begin with Descartes' arguments for dualism of mind and body—the view that minds and bodies are distinct and incompatible kinds of substance, *res cogitans* and *res extensa*. Most philosophers now regard Descartes' arguments to be defective, and even some of his contemporaries saw clearly the main problems that his dualism faced. Nevertheless his construction of the problem of understanding the mind remains influential. Doing better than Descartes' dualism is a benchmark for philosophical theories of the mind. And, for many philosophers of mind, the fear that one's view collapses into dualism—or even resembles it—remains a serious one.

It is time to move beyond this archaic framework. As Jaegwon Kim correctly notes, since the mid-twentieth century "the mind-body problem—our mind-body problem—has been that of finding a place for the mind in a world that is fundamentally physical" (1998: 2). But rejecting Cartesian dualism is one thing; finding a compelling alternative quite another. The correct response must do more than simply narrow serious proposals about the nature of mind to those that admit only physicalist monism. The lesson of dualism's failure is not just that Descartes was wrong that minds are *res cogitans*. Rather, the lesson is that the study of the mind must proceed with the same conceptual tools and resources that philosophers and scientists have applied in all other natural domains. The right response is not simply to restrict the possible space of answers, but to reframe the question. We take as a starting point that a science of the mind should begin—and will someday end—in the same world of mountains, seas, animals, and atoms that other sciences investigate. As Kim emphasizes, we must pursue "mind in a physical world." In doing so, we finally abandon the old mind-body problem once and for all and replace it with the more familiar, but of course still difficult, problem of developing the broadly empirical study of minds.

Some philosophers worry that adopting an empirical approach to the nature of minds threatens the significance of their contributions. Just as a kind word and a couple of dollars will get you a cup of coffee, so too a philosopher and a neuroscientist will crack the puzzle of mentation. But if the worry is that philosophers will not always have the last word about the properties of minds, then we see no great loss. On the other hand, if we understand philosophy as trying, as Wilfrid Sellars suggests, "to understand how things in the broadest possible sense of the term hang together in the broadest possible sense of the term" (1963: 1), then philosophers need not worry that there is a choice to be made between scientific and philosophical theorizing. That is another anachronistic dualism that we do well to leave behind. Understanding what it means for minds to be physical, and the implications of this claim for the sciences of the mind, leaves plenty to do for philosophers, even granting the need of empirical methods for a complete picture of mental phenomena.

Acknowledgments

We would like to thank audiences in Madison, Shreveport, Cincinnati, Birmingham, Dover, Berlin, Lausanne, Paris, Jerusalem, Cologne, Leuven, Kazimierz Dolny, Seoul, Evanston, the Philosophy of Science Association, the Society for the Metaphysics of Science, the Southern Society for Philosophy and Psychology, and the International Society for the History, Philosophy, and Social Studies of Biology.

Our view of multiple realization developed incrementally at a series of workshops over the past decade, whose attendees included: Fred Adams, Ken Aizawa, John Bickle, Dan Brooks, Cédric Brun, Carl Craver, Zoe Drayson, Markus Eronen, Carrie Figdor, Jörg Fingerhut, Gary Fuller, Carl Gillett, Jens Harbecke, Vera Hoffman-Kloss, Tobias Huber, Philippe Huneman, Marie Kaiser, Jesper Kallestrup, Lena Kästner, Beate Krickel, Corey Maley, Alex Manafu, Mohan Matthen, Karen Neander, Marieke Rhode, Bob Richardson, Rob Rupert, Christian Sachse, Raphael Scholl, Elliott Sober, Patrice Soom, Jackie Sullivan, Kari Theurer, Sven Walter, and Markus Wild. We have also benefited from feedback from research groups based in Cologne and Jerusalem that include some of the above people as well as Eli Dresner, Meir Hemmo, Arnon Levy, Oron Shagrir, and Orly Shenker.

We are especially grateful to our friends and colleagues who discussed portions of this manuscript with us: Ken Aizawa, Rosa Cao, Tony Chemero, Valerie Hardcastle, Doug Keaton, Brian Keeley, Colin Klein, Gualtiero Piccinini, Angela Potochnik, Bob Richardson, Christian Sachse, Rob Skipper, and Elliott Sober. We are also indebted to Peter Momtchiloff, Eleanor Collins, Sarah Parker, Dawn Preston, and two anonymous readers for Oxford University Press for shepherding this project.

Tom Polger's work on this project was made possible in large part by support from the Charles P. Taft Fund and Taft Research Center at the University of Cincinnati, and by a sabbatical leave from the University of Cincinnati. Larry Shapiro is grateful to the University of Wisconsin for a sabbatical leave during 2014–15.

List of Figures and Tables

PART I

Whence
Multiple Realization?

1

Physicalism and Multiple Realization

1. Post-Cartesian Physicalism
2. Functionalism and Explanation in Psychology
3. What Good Is Multiple Realization?

1 Post-Cartesian Physicalism

Once we stop worrying about Descartes' mind-body problem and start worrying about the place of the mind in a physical world, we must set about evaluating theories about minds in the same ways that we evaluate broadly scientific theories of anything else. In short, we must view philosophy of mind as a species of philosophy of science. And theory selection in the philosophy of mind will be like theory selection in any other broadly scientific inquiry.

The pioneering twentieth-century physicalists who theorized about the mind defended their views precisely within the context of such broadly empirical considerations. Criteria for theory selection that have been especially salient for philosophy of mind-as-science include empirical adequacy, simplicity, relevance, and generality. Herbert Feigl, U. T. Place, and J. J. C. Smart articulated and defended a mind-body identity theory—the view, as Smart put it, that "sensations are brain processes" (1959: 144, 1961). Already by the mid-1900s, all three believed that the empirical evidence justified the belief that a brain-based theory of psychology would be empirically adequate. And, although they differed on precisely how to apply the criterion of simplicity—a vexed notion in its own right—they each held that the mind-brain identity theory is simpler than the alternatives. In particular, they argued that the identity theory is simpler than dualism, which at that time remained among the main competitors.

But it is important to note that the early identity theorists didn't argue that the identity theory is simpler than dualism merely because the identity theory postulates only one kind of stuff—the physical stuff— whereas dualism postulates two. That application of simplicity would cut no ice if the dualist theory put the extra mind-stuff to good work. Rather, in light of the difficulty of understanding how mind-stuff could causally interact with physical stuff, the early identity theorists concluded than any version of dualism would have to endorse *epiphenomenalism*. The dualist, that is, would be forced to concede the causal impotence of mind-stuff. Mind as distinct from matter would make no difference in the world. So the complaint against dualism is not just that it postulates two kinds of stuff, but that it postulates a kind of stuff that plays no role in explaining the capacities of creatures who have minds. A main purpose of theorizing about minds, after all, is to construct an explanation of the behaviors, thoughts, feelings, and so forth of creatures such as us. Epiphenomenalist theories are less simple because they postulate entities that don't appear to contribute to that explanatory project. In applying the theoretical criterion of simplicity in this way, the early identity theorists were not just counting substances but endorsing more substantive desiderata: Psychological states, whatever they are, should have some causal effects.

In addition to simplicity, some of the identity theorists also emphasized another theoretical virtue that they saw as speaking in favor of their own view and against Cartesian dualism. Smart asserts "That everything should be explicable in terms of physics . . . except the occurrence of sensations seems to me to be frankly unbelievable" (1959: 142). Here Smart expresses a kind of conservatism. Every successful scientific endeavor—physics, geology, chemistry, biology—adopts a purely physicalist perspective, at least in the sense that the domains they investigate are assumed to consist, ultimately, of physical matter. One should expect for this reason that the study of minds would conserve this trend. Certainly, in any case, one should pursue a science of the mind with the assumption that its main commitments will be consistent with those of other sciences rather, as the Cartesian dualist maintains, wholly distinct.

Another mid-twentieth-century approach to the mind responded to the perceived consequences of dualism differently. Behaviorists argued that postulating "internal" psychological states—whether states of brains or of mind-stuff—is not necessary for understanding the operations of the mind. Psychology ought simply to be, as John Watson put it, "the science of behavior" (1913). Internal psychological states are as irrelevant for explaining behavior as immaterial thinking substances would be. B. F. Skinner, for example, argued adamantly that internalist explanations of psychology, whether mentalistic or neurological, adopt indefensible and unnecessary theoretical posits (1953). From here, it is a short leap to the idea that internal psychological states do not exist at all. Some behaviorists thought that the body is just a behavior machine, as might be true of organisms like wasps and whelks (cf. Dennett 1984 and Keijzer 2013). Psychological states, if mentioned at all, are treated as states of the total organism, rather than internal causes or working parts.

By the late 1960s it seemed clear that behaviorism, at least in its sparest forms, could not meet the minimum standard of empirical adequacy. Chomsky argued that its predictions did not match the data, or that the behaviorist program rested on a vicious sort of post-hoc reasoning (1959); and his criticisms convinced many philosophers and cognitive scientists.[1] But even before Chomsky, behaviorism's opponents noted that the use of simplicity to argue against all "internal" theories of the mind is problematic. Simplicity is only one criterion among many, and one that is decisive (if at all) only when all other factors are equal. The behaviorist does not have a clear case for simplicity, but at best a case of competing theoretical values.

Nonetheless, despite behaviorism's failures, we might still find reason to praise its commitment to the same kind of conservatism we see Smart embracing. As the identity theorists before them, behaviorists resisted the desire to characterize mental phenomena as *sui generis* and inexplicably apart from the rest of the natural world that scientists hope to describe and understand. If in the end they denied the existence of the very phenomena that we might suppose should be the target of psychological investigation, they did so not because these phenomena were beyond the reach of a physically oriented psychology, but because, so they believed, such a psychology could proceed without them.

---

[1] Chomsky's influence is generally recognized even by those who are critical of his arguments, e.g. MacQuorcodale (1970) and Palmer (2006).

This cartoonish history of the debate in mid-twentieth-century philosophy of mind and psychology doesn't do justice to any of the views, to the nuances of the debate, nor to the serious scientific work that has been conducted within each theoretical framework— behaviorism's methodological legacy for the mind sciences, in particular, is not to be underestimated. We relate this history not with goals of completeness or exactness, but with the explicit agenda of highlighting how theorists on all sides appealed to broadly scientific and theoretical considerations in advancing their views. This reconfiguration of the framework for the inquiry, more than a mere rejection of Cartesian dualism as an answer to questions about the mind, is the important advance pushed by identity theorists and behaviorists alike.

2 Functionalism and Explanation in Psychology

The advent of modern computing technologies provided a concrete model for the old metaphor that minds are calculating machines. Throughout the 1950s Hilary Putnam, along with many others working in philosophy and in psychology, explored the parallels between psychological and computational systems. Putnam advanced his functionalist hypothesis within the context of this new empirical approach to understanding minds. The idea that thinking is a kind of information processing was not new. Electronic computers, however, provided a whole new conception of what an information processor might be. Mechanical calculators had been around at least since the abacus. But the kinds of machines relevant to modern electronic digital computing do not have levers, pulleys, or gears—components obviously not present in the brain. Moreover, the central notion of a computational state seemed, to Putnam, to apply as well to states of the brain.

So, Putnam's innovation was not to suggest that thinking is like calculating, or that minds are like computing machines. Rather, his advance lay in proposing the literal identification of minds with classes of computing machines, suitably understood. And he advanced this theory not as a philosophical thesis, but as an empirical hypothesis: "I shall not apologize for advancing an empirical hypothesis . . . I propose the hypothesis that pain, or the state of being in pain, is a functional state of the whole organism" (1967 in 1975: 433). Putnam explained the idea of a "functional state of a whole organism" in terms of the states of a particular kind of computing machine, a probabilistic automaton. His initial functionalist hypothesis claimed simply that psychological states are such states, and accepted the implication that minds are just such machines.

Putnam defended his hypothesis against the two alternatives mentioned above: Behaviorism and the identity theory. Against behaviorism, Putnam rehearsed some well-known objections to the theory that was, by that time, already on the ropes. He remarks, "difficulties with 'behavior disposition' accounts are so well known that I shall do little more than recall them here" (1967 in 1975: 438). But against the mind-brain identity theory, Putnam forcefully developed a novel line of thought. He argued that the functionalist hypothesis is both simpler and more general than the mind-brain identity theory (Putnam 1967 in 1975: 436–7). Putnam begins with the observation that a diversity of biological creatures might share some psychological states. Indeed, the neutrality of the functionalist hypothesis leads him to suggest that even non-physical systems might instantiate the same psychological states as physical ones (1967 in 1975: 436).

Because psychological states are *multiply realized*—as the phenomenon came to be called—the functionalist theory exhibits more generality than any theory that identifies mental states with the states of particular kinds of brains. More perspicuously, any theory that identifies psychological states with the states of particular kinds of brains will be less general than a theory, such as functionalism, that identifies psychological states with "abstract" computing states that that can occur in various kinds of brains (Putnam 1967 in 1975: 436).

Moreover, according to Putnam the functionalist theory is simpler than the identity theory because it will include species-independent psychological generalizations. In contrast, he speculates that the only plausible way that a mind-brain identity theory could even potentially match the generality of functionalism would be to postulate sets of species-specific or brain-type specific laws. Such a maneuver could succeed in unifying psychological states only by introducing kinds of a very unusual sort. The kind pain, for instance, is reimagined as the open-ended set of {*pain-in-humans* or *pain-in-dogs* or *pain-in-octopuses* or . . . }. Although this, he conceded, would allow the identity theorist to give an account of pain, it would work only by invoking the artificial (and arguably ad-hoc) device of disjunctive kinds. On the other hand, functionalism achieves generality at no extra cost, doing so with fewer and simpler laws than the identity theory would require (Putnam 1967 in 1975: 437).

When Putnam introduced the multiple realization considerations in favor of functionalism, he did so with the confidence that they would eventually win empirical support. Some years later he advanced a more normative argument that we should expect explanations in psychology, and in non-fundamental sciences in general, to be broadly functional. He writes:

People are worried that we may be debunked, that our behavior may be exposed as really explained by something mechanical. Not, to be sure, mechanical in the old sense of cogs and pulleys, but in the new sense of electricity and magnetism and quantum chemistry and so forth . . . part of what I want to do is to argue that this can't happen. Mentality is a real and autonomous feature of our world. (1973 in 1975: 291)

Here Putnam describes a concern about what he calls the "autonomy" of psychology, of our mental life. The concern is that the explanation of our mental life—our psychology—in terms of anything non-psychological, would have the consequence of deflating or undermining psychological explanation, and consequently of undermining our status as psychological beings.

But Putnam goes further than simply denying that his functionalist theory of mind entails a loss of psychology's autonomy. He claims, moreover, that the autonomy of psychology is assured regardless of whatever mechanistic theory of mind eventually proves to be true. Putnam defends this conclusion by means of his famous example of a square peg and a board with a round hole whose diameter equals the width of the peg. Putnam argues that there is no good micro-based explanation for the peg's failure to fit through the hole. "One could compute all possible trajectories" of the peg's molecules, he says, "and perhaps one could deduce from just the laws of particle mechanics or quantum electrodynamics" that the square peg never passes through the round hole of the same diameter (1973 in 1975: 295). But, Putnam urges, although a micro-based calculation of the systems' interactions in terms of the trajectories of particles might be possible, such a calculation would not explain what is going on in the system. How should it be explained? "The explanation is that the board is rigid, the peg is rigid, and as a matter of geometrical fact, the round hole is smaller than the peg . . . That is the correct explanation whether the peg consists of molecules, or continuous rigid substance, or whatever" (1973 in 1975: 296). In contrast, Putnam says that the micro-based explanation is either "not an explanation" or "just a terrible explanation" (1973 in 1975: 296).[2] The macro-geometrical explanation is preferable, Putnam insists, because it picks out the relevant factors (viz., the geometric ones) and leaves out irrelevant factors (e.g., the material composition of the peg and board.) Furthermore, it is more general because it applies to pegs and boards no matter what they're made of. Though he admits that the micro-based explanation generalizes in a different way, Putnam denies that this marks an explanatorily valuable kind of generality.

Accepting that a functionalist description of mental states stands as the analog to the macro-geometrical description of the peg, the lesson that Putnam draws is not just that his functionalist theory is superior to the identity theory as an empirical hypothesis, but moreover that scientific psychological theories ought to employ just such functional-psychological kinds because they invariably have greater explanatory relevance and generality.[3]

The canonical form of this line of thought comes from Jerry Fodor. In an influential article in which he advanced something like the argument for the non-reductive or realization theories that we are discussing here, Fodor writes:

The reason it is unlikely that every natural kind corresponds to a physical natural kind is just that (a) interesting generalizations (e.g., counter-factual supporting generalizations) can often be made about events whose physical descriptions have nothing in common, (b) it is often the case that *whether* the physical

---

[2] But see Sober (1999) for a dissenting view.
[3] Here a historical note and a philosophical note are merited. The historical note is that in this paper Putnam means to argue against his previous functionalist hypothesis: But this is not (yet) because he rejects functionalism in general, only because he rejects what he understands to be a narrow conception of function in terms of Turing machines. He says that "functional organization is extremely important," but that notion "became clear to us through systems with a very restricted, very specific functional organization" (1973, in 1975: 300). The philosophical note is that one should not be tricked into thinking that this "autonomy" argument does not depend on multiple realization considerations: It is only because Putnam assumes that pegs and boards are multiply realized that he can argue that explanations in terms of them have different relevance and generality than the micro-based explanations with which he contrasts them.

descriptions of the events subsumed by these generalizations have anything in common is, in an obvious sense, entirely irrelevant to the truth of the generalizations, or to their interestingness, or to their degree of confirmation or, indeed, to any of the epistemologically important properties, and (c) the special sciences are very much in the business of making generalizations of this kind. (1974: 103)

After asserting that the special sciences are "very much" in the business of making generalizations for which information about the things on which they depend is irrelevant, Fodor offers the example of Gresham's law in economics to illustrate the point. The law describes a pattern in the proliferation of kinds of monetary currency, "good money drives out bad." But—and here enters the significance of multiple realization—the law remains true regardless of the material composition of the currency. The idea, then, is that we have sciences like psychology because there are laws or regularities that they capture but that are not laws or regularities in more basic sciences because the "higher level" kind is multiply realizable by heterogeneous "lower level" kinds (Figure 1.1).

Later we will return to the famous passage from Fodor, above. For now we note that the idea that psychological kinds are functional kinds is no longer, in Fodor's hands, an ordinary empirical hypothesis about mental states, intended as a competitor to other empirical hypotheses such as the identity theory or behaviorism. Rather, we are told, we should expect that the "special sciences"—the sciences that describe the behavior of non-fundamentally physical kinds—will concern functional kinds. Fodor offers little in the way of argument for this claim, commenting only that "these remarks are obvious to the point of self-certification; they leap to the eye as soon as one makes the (apparently radical) move of taking the special sciences at all seriously" (1974: 103).

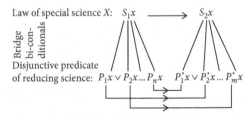

Figure 1.1 Fodor's picture of the special sciences. From J. Fodor (1974). With permission of Springer.

Our focus, like Fodor's, is on the cognitive and brain sciences. But it is worth noting that the justification of the special sciences by the multiple realizability of their kinds is the dominant way of thinking about the non-fundamental sciences among philosophers—even, or especially, those outside the philosophy of science. So understanding multiple realization is important even for those who are not concerned about minds and brains.

With the backing of Putnam, Fodor, and others, by the early 1970s realization theories under the guise of functionalism had become the dominant theories in the philosophy of mind. Subsequent theorists, including Putnam himself, elaborated the functionalist hypothesis in a variety of ways.[4] The framework of a machine program table was replaced by that of an empirical or analytic psychological theory, with the theory specifying the psychological states and state transitions. And computational relations have by and large been replaced by causal relations as the constitutive essences of functional states. Sometimes the functions with which psychological states are identified are required to have evolved in particular ways (Fodor 1968, Lycan 1987, Sober 1990). All of these views hold in common the idea that mental states are physical states insofar as they are *realized* or *implemented* by physical states, just as computing machines inherit their physical credentials from the hardware in which they are realized or implemented.

Some contemporary philosophers reserve "functionalism" as a label for the old view that minds are computing machines, and advocate instead a purportedly non-functionalist *realization theory*. These theorists hope to take advantage of the resources of the idea that mental states are realized by physical states, but without some of the baggage that they associate with functionalism.[5] However, when the question arises as to why functionalist or realization physicalist theories (broadly, *non-reductionist* theories) are to be preferred over brain-based and identity theories (broadly, *reductionist* theories) inevitably it is the multiple realization of the mental that is cited. As multiple realization is our

---

[4] Jerry Fodor (1974), David Lewis (1970, 1972), Jaegwon Kim (1972), Kitcher (1984), Bill Lycan (1987), Putnam (1988, 1999), and Ned Block (1978, 1980), among others. See Polger (2004a, 2008, 2012) for discussion of different versions of functionalism.

[5] Putnam himself abandoned first the details of his 1967 proposal (1973 in 1975), and then the general spirit of the view (1988, 1999).

concern, we will often speak indifferently of "functional" and "realization" theories.[6]

We will take for granted that all functionalists and realization theorists
hold that psychological states are at least possibly multiply realized—that
is, they are at least *multiply realizable*. Of course, there may be reasons
for being attracted to a functionalist or realizationist theory that have
nothing to do with multiple realizability. Nevertheless, our present
concern is not whether there are some positive reasons for adopting
functionalism. Rather we are specifically considering whether multiple
realizability provides a reason to prefer functionalist or realizationist
theories over the mind-brain identity theory.

Seemingly overnight Putnam's multiple realization argument convinced
philosophers that the mind-brain identity theory is untenable, and that his
functionalist alternative was preferable. Moreover, as it became clear that
the details of his original hypothesis were indefensible—mental states
are not Turing machine states—the theory only became more popular.
For Putnam's empirical hypothesis about "the nature of mental states"
morphed into a general framework for understanding various non-
fundamental features of our world. Putnam, then Fodor, David Lewis,
Philip Kitcher, William Lycan, Frank Jackson, and Philip Pettit, and many
others extended the functionalist theory from psychology to a variety of
other sciences and beyond. In every case, multiple realization is cited as
evidence for "non-reductive" realization theories in contrast to "reductive"
identity theories. We will be focusing our attention on the original formu-
lations of the multiple realization arguments and on the accompanying
vision of how to justify explanation in the non-basic sciences. Though
seminal works of Putnam and Fodor are now approaching their silver
anniversary, the views they advanced are both currently popular and
thoroughly entrenched. Multiple realization seems so obvious to many
philosophers of mind and science that arguments in its favor are rarely
demanded, and none have been given that do not hark back to those
founding documents. So they will be our touchstones.

In the meantime, the core idea that many non-fundamental features
are realized by but not identical to fundamental or more fundamental
features has now advanced outside of the philosophy of science to ethics,

---

[6] We also think that grouping them together is accurate, that they are a unified family of
theories.

metaphysics, epistemology, and metametaphysics. In every case, the arguments for functional or realization theories—whether of *good* (cf. Smith 1994, Jackson and Pettit 1995), *yellow* (cf. many theorists), *truth* (cf. Lynch 2000), or general ontology (cf. Lewis 1994, Jackson 1998, Melnyk 2003)—depend on the claim that the non-fundamental entities are "nothing over and above" more fundamental things doing a certain job or function, which function could be performed by something else. That is, multiple realization enjoys a central role in a wide swath of theories in philosophy of science, metaethics, metaphysics, and epistemology. And in every case they take their cue from the apparent—we think, usurped—success of realization theories in philosophy of psychology and mind. For this reason, the significance of multiple realization cannot be overestimated, and assessing the case for multiple realization is of pressing importance. Of course, if multiple realization in philosophy of psychology is not what it is cracked up to be, other realization theories might still succeed. But if the paradigm case fails, we expect that the ripples of this failure might upset many philosophical domains.

3 What Good Is Multiple Realization?

As we are setting out the dispute, the argument that a non-reductive theory of psychological states (viz., functionalism) is preferable to a reductive theory (viz., the identity theory, or, as we'll also refer to it, the brain state theory)[7] depends on the premise that mental states are multiply realized, or at least multiply realizable. Putnam argues, on these grounds, that the functionalist theory is both more general and simpler than the brain state theory. This is because insofar as the brain state theory matches the generality of the functionalist theory, it will do so only at the cost of simplicity. Must that be the case?

Putnam thinks so, because any psychological theory that fails to cover the full range of known psychological creatures will, he believes, fail to be empirically adequate. And Putnam holds that mental states are not just multiply *realizable*, but that they are in fact multiply *realized*. In a

---

[7] We use the term "reductive" with some reluctance, because the philosophical world teems with theories of reduction, some of which might be compatible with multiple realization and some of which are not commitments of our identity theory. In fact, we will defend an explicitly pluralist approach in Chapter 10.

famous passage, he explains why the brain state theory strikes him as incredibly implausible: "the physical-chemical state in question must be a possible state of a mammalian brain, a reptilian brain, a mollusc's brain (octopuses are mollusca, and certainly feel pain), etc. At the same time, it must *not* be a possible (physically possible) state of the brain of any physically possible creature that cannot feel pain" (1967 in 1975: 436). Putnam simply dismisses the possibility that there could be neuro-psychological laws that are general enough to span mammals, reptiles, and molluscs but also not so general as to include some physically possible creature that cannot feel pain. He says, "It is not altogether impossible that such a state will be found... But this is certainly an ambitious hypothesis" (1967 in 1975: 436).

In effect, Putnam challenges the brain state theorist with a nested dilemma. Either the brain state theory attempts to give an account of the same range of phenomena as the functionalist theory, or it does not. If not, then the functionalist theory wins points for generality. If so, then another dilemma appears: Either the brain state identity theory can give unified accounts of the pains of mammals, reptiles, and molluscs or it cannot. If it cannot, then it will be empirically inadequate. If it can, it must do so through the introduction of a set of species-specific neuro-psychological regularities—but that is to compromise explanatory simplicity. Subsequent theorists have worried that saving generality in this way may also amount to denying the reality of the psychological states in question (cf. Block 1997, David 1997, Aizawa and Gillett 2011). Putnam himself suggests this worry when he introduces multiple real-ization, and again when he offers the example of the square peg and the round hole. For if "pain in humans" and "pain in octopuses" are not instances of the same property, then in some sense we have eliminated *pain*, as such, from our psychological theory. And even if the eliminati-vist slide can be blocked, the resulting theory will still lack simplicity. Thus even if the brain state theory is empirically adequate, it remains less simple than the functionalist theory. Putnam concludes that functional-ism is either more general, more empirically adequate, or simpler than the identity theory.

Both identity and realization theories, it is admitted, have certain advantages over one or both of Cartesian dualism and behaviorism. Compared to dualism, both the realization and identity theories appar-ently have ready explanations of the causal efficacy of mental states, in terms of the causal efficacy of the physical states that realize them or are identical to them, respectively. (Though whether realization theories can ultimately vindicate mental causation is notoriously controversial. We will discuss mental causation in Chapter 10.) And, compared to behaviorism, both theories are realist about mental states. Perhaps that is even a necessary condition on psychological states having causal powers; but it is certainly not sufficient, as epiphenomenalist dualism postulates real mental states that lack causal efficacy.[8]

At this point, then, we can see the contemporary dispute between the realization theorists and the identity theorists as one over which theory better satisfies a relatively well understood—though not always explicitly articulated—set of desiderata. The first two desiderata we have already encountered:

(R)   Realism. Psychological states are genuine internal states of systems that have them, *pace* behaviorism.
(C)   Causal efficacy. Psychological states have causal powers, *pace* epiphenomenalist dualism.

The third desideratum is the one pressed by Putnam's multiple realization argument:

(G)   Generality. Psychological states are general rather than species-bound, such that a range of creatures may possess the same psychological state.

Inability to accommodate this fact is supposed to be the fatal flaw with the brain state theory.

Additionally, we know that because the realization and identity theories are put forth as empirical hypotheses, each is supposed to be empirically adequate. But to what empirical data are the theories to be adequate? One bit of that data is the extent of correlation or lack of correlation between psychological and neurological kinds. Insofar as identity and realization theories are broadly empirical theories about minds, they must be adequate to discoveries of such correlations. This requirement is partly reflected in the above desiderata that psychological

---

[8] Jaegwon Kim (1998) argues that we should adopt "Alexander's Dictum" that to be real is to have causal powers. In that case, epiphenomenalist dualism is not realist about mental states, after all.

state kinds not be species-specific; not to mention the requirement for realism about the mental.

But in addition, as philosophical or meta-theories, they must also be adequate to the broadly empirical information we have about the sciences of the mind themselves. Both realization and identity theorists claim that their theories accommodate the explanatory practices of the sciences of the mind—neurobiology, neuropsychology, psychology, cognitive science, and the like.[9] Indeed, many aspects of these various sciences come into contact with the philosophical theories, some of which we shall consider in later chapters. But one deserves highlighting immediately. Namely, any philosophical theory of the nature of minds must recognize a legitimate role for psychological explanation, that is for explanation in which the *explanans* includes one or more psychological states or events:

(E)   Explanatory Psychology. Psychological states can figure in legitimate psychological explanations.

Sometimes this aspect of the desideratum is put in terms of the requirement to vindicate the "autonomy" of psychology. We shall return to the vexing issue of whether psychology is autonomous in some way from the brain sciences in Chapter 10.

Once we have a set of desiderata like these, we can check to see how the competing theories fare. Do they account for the Reality and Causal efficacy of mental states? Do they account for the Generality and Explanatory legitimacy of psychological explanations? We take it that, according to functionalists and other advocates of realization theories, the scorecard for the competing theories of mind should provisionally look something like Table 1.1.[10]

According to this way of counting, the identity theory is an attractive theory, except that—crucially—it fails to be as general as the functionalist

---

[9] There are those who hold that the sciences of the mind should be constrained by our folk psychological theories. One could read David Lewis' proposals in this way (1970, 1994). What little we have to say about that idea will be said in Chapter 4 and then in Chapter 9 when we discuss the problem that we call "Putnam's Revenge."

[10] Note, especially with respect to the question of explanations, the checkmarks reflect rather different claims in the dualist, behaviorist, or functionalist frameworks. Behaviorist mental states are states of the whole organism rather than either internal or "psychical" states, for example.

Table 1.1. Scorekeeping theories in the philosophy of psychology

	Real	Causal	General	Explanatory
Dualism	✓		✓	
Behaviorism			✓	✓
Identity Theory	✓	✓		✓
Realization	✓	✓	✓	✓

or realization theory. That is, all other theories are compatible with psychological processes occurring in creatures—or other systems—that are quite unlike human beings.[11] For example, as Bill Lycan sees it, the identity theory is "an empirically special case of Functionalism ... that (implausibly) locates all mental states at the same very low level of institutional abstraction—the neuroanatomical" (1987: 59).

Importantly, functionalism's main virtues trace to the supposed physical realization and, moreover, *multiple* physical realization of mental states and processes. Because they are physically realized by brain states, psychological states are real. But because they are multiply realizable, they cannot be identified with any particular brain state. On the flip side, their multiple realizability explains the generality of psychological phenomena—many different creatures can have pain, hunger, belief, and so forth. Furthermore, if psychology is the science of these multiply realizable functional states, then there exists a legitimate explanatory role for psychological sciences beyond that of the brain sciences. So the claim that mental states are realized is quite important to functionalism, or any realization theory.

It is thus crucial to the success of functionalism, and realization theories generally, to determine whether psychological states are in fact realized. And, in order to determine that, one must discover whether they are in fact multiply realized or even multiply realizable. This question brings us to the next chapter.

---

[11] We count substance dualism as general because, although immaterial substance is required, we assume that thinking substance could be "superadded" to any other system— Descartes' and Locke's qualms notwithstanding. But we also suppose that someone who thinks that dualism is incompatible with mental causation (as we do) will also and for the same reasons think that dualism cannot vindicate a science of psychology. That latter assessment is controversial, and we will not defend it herein.

2

Realization and Multiple Realization

1 What Is Realization?

If we are going to determine whether psychological states are, in fact, multiply realized or multiply realizable, then we're going to need a more precise understanding of what the phenomenon of multiple realization involves—and also of what the realization relation might be. As we have situated these questions, potential answers must be judged against their ability to do the work needed in the line of reasoning sketched in the previous chapter. That is, the fact that psychological states are realized must secure their Realist credentials in a way compatible with their multiple realization. And the fact that psychological states are multiply realized must guarantee their Generality. And both of these facts must be compatible with the Causal potency of psychological states, and with the Explanatory utility of the sciences that explain and explain-by-means-of those states.

The non-reductive physicalist claims that psychological state kinds are realized by, but not identical to, brain state kinds. Is the qualification "but not identical to" intended to suggest the possibility of psychological states that might be realized by *and also identical to* brain states? We think not. As we understand the realizationist view, the claim that psychological states are realized by "but not identical to" brain states is an elaboration of the idea of realization rather than an additional claim. That is, if mental states are realized by brain states then they are *ipso facto* not identical to brain states.[1] The identity of psychological states with brain states is an incompatible alternative to their being realized by brain states.

Sometimes the identity theory and behaviorism are portrayed as two poles of a continuum, with functionalism occupying the middle ground. For example, Ned Block's problem of inputs and outputs is framed in terms of trying to find an intermediate way of characterizing psychological states (1978). The solution is for them to be sufficiently abstract to avoid the extreme "chauvinism" of insufficiently general identity theories, but also not so abstract that they overgeneralize as extremely "liberal" behaviorist theories are prone to do (1978 in 1980a: 293–6). Thinking of the three alternatives on a continuum is convenient for certain purposes, especially for understanding the motives behind functionalism. But if we are right, it misleadingly implies that identity theory and behaviorism are just extreme versions of functionalist views. This is a mistake; nothing can be both realized by and identical to any one thing. Identity is not a special or limiting case of realization.[2] So, as we shall further explain, an account of realization should discriminate between those states or processes that are realized, e.g., by brain states or processes, and those that are not.

What is realization? To begin, realization is a metaphysical dependence relation. According to the advocates of non-reductive physicalism, realization is the metaphysical dependence relation between brains and minds. We will not, here, try to say anything substantial about what metaphysical dependence itself might be. For our purposes, the important idea is that the realization of mental states by brain states must be such that it, as the ordinary connotation of "realization" suggests, "makes real" those mental states. Thus the kind of metaphysical dependence that we have in mind is sometimes called *ontological dependence*, according to which the dependent thing would not exist (or would not exist as the

---

[1] Realization theories are sometimes said to be "token identity" theories, as each mental token is identical to some physical token (Davidson 1970). We agree with Kim (2012) that token identity entails type identity. But at any rate, it is type identity—the identities of kinds—with which we are concerned. When we speak of the identity theory, we mean the type-identity theory.

[2] So on our view, Lycan is wrong to claim, as we noted in the previous chapter, that the identity theory is just "an empirically special case of Functionalism . . . that (implausibly) locates all mental states at the same very low level of institutional abstraction—the neuro-anatomical" (1987: 59).

kind of thing that it is) without the existence of that on which it depends. For example, a desk would not exist if the desktop and legs did not exist—the existence of the desk depends on the existence of its parts, the desktop and legs. And Van Gogh's painting, *Undergrowth with Two Figures*, would not exist without some paint being arranged on a canvas in a certain way—the existence of a painting depends on the existence of the paint.[3]

We're going to help ourselves to the idea of ontological dependence or "building" relations (Bennett 2011, forthcoming). Furthermore, we think it is widely agreed that the kinds of ontological dependence at stake for the understanding of realization and multiple realization are synchronic and constant—current psychological states ontologically depend on current brain states, and do so continually for as long as they persist. Of course they also depend on past brain states. Causation is a non-synchronic ontological dependence relation; effects ontologically depend on their preceding causes. Van Gogh's painting would not exist without the paint, but it also would not exist had Van Gogh himself not existed. But the functionalist or realization theorist does not merely claim that psychological states depend on brain states historically or non-synchronically, e.g., by being caused by them. Rather, the claim is that psychological states constantly and continuously depend on brain states. Psychological states are not realized by brain states merely in the way that Van Gogh "realized" his paintings by painting them. Rather, psychological states depend on brain states in something more like the way that paintings continue to depend on the paints of which they were painted. Remove the paint and you've removed the painting.[4]

Prima facie there are a variety of ontological dependence relations. For instance, *composition* is the relation between my desk and its parts.

---

[3] Examples of dependent entities include particulars, properties, and kinds. We will have something to say about the relata of the realization relation shortly.

For accounts of ontological dependence, see: Fine (1995, 2001), Thomasson (1999), Bennett (2011). There is a long history of trying to analyze these dependence relations in terms of supervenience relations, and it is likely that dependent entities do supervene on their bases. But there is now ample reason to think that dependence is not merely supervenience (e.g. Horgan 1993a, Kim 1998, Fine 1995, 2001).

[4] We take this example at face value—we are talking about the paintings as particular objects. We take no stand on the ontology of "works" of art, for example on whether *Undergrowth with Two Figures* would still exist if the painting were destroyed but photographs of the painting survived.

Constitution is the relation between a statue and the bronze out of which it is made. *Identity* is the relation between, e.g., Mark Twain and Samuel Clemens, or gold and the substance with the atomic number 79. And prima facie there is *realization*—the ontological dependence relation between computing machines and physical devices; and, if non-reductive physicalists are correct, between psychological states and brain states.

Although we shall not attempt to provide an account of ontological dependence in general, we can say some useful things about the sort of ontological dependence relation that is realization. As we have just seen, it is synchronic and constant. Whatever is realized depends continuously on the existence of the realizers on which it depends. Second, also noted above, realization is incompatible with identity. What is realized is not identical to its realizer or realizers.

Third, realization must be a dependence relation that transmits physical legitimacy from realizers to what they realize. Being physically realized must be sufficient for being physical, broadly speaking. If being physically realized did not ensure the physicality of what is realized, then the non-reductive theory could not be guaranteed to be a physicalist theory after all. Now, it need not—indeed should not—be the case that realization is a "physicalizing" relation, such that whatever is realized is physical. Putnam was clear that there could be, as far as he was concerned, non-physical realizers of functional states—they might be made of "copper, cheese, or soul," he said (Putnam 1973 in 1975: 292). We post-Cartesian physicalists don't worry much about immaterial thinking substances turning up as realizers. But what is quite crucial for the non-reductive physicalist is that if something has a physical realizer then it inherits the physical credentials of its realizer.

Finally, realization must be compatible with multiple realization. After all, Putnam introduced realization as a metaphysical dependence relation between brains and minds that is compatible with the multiple realizability of psychological states. If realization were not at least compatible with multiple realization, then realization theories would not have the advantage of generality over other theories, and specifically would not have that advantage over the mind-brain identity theory.

What, then, is the relationship between realization and multiple realization? It is tempting to think that to understand multiple realization we must first understand realization. But this gets it exactly backward (Polger 2015). To understand realization we must first understand the empirical phenomenon of multiple realization. Simply put, realization is the postulated ontological dependence relation between brains and minds that is compatible with the multiple realizability of psychological states. This means that the term "multiple realization" is a bit misleading. (In fact, Putnam did not use that term in his original arguments.) It would be better to say that realization is the ontological dependence relation between brains and minds that captures the postulated Generality of psychological states—the expected extent of physical variation among the various creatures that have psychological states.

So, realization is a synchronic ontological dependence relation, distinct from identity, and that transmits physical legitimacy from physical realizers to what is realized. And, finally, whatever is realized must be multiply realizable. What sort of relation satisfies these desiderata? We prefer to understand the relation of realization in terms of the relation of having a function:

Realization: Necessarily, P realizes G if and only if P has the function F_G.

To realize G is to have the function constitutive of G's, the G-function. This formulation is a schema that could be filled in with any of various notions of function, with only the constraint that identity is not a realization relation. The idea is just that for some entities—properties, states, kinds, objects—*being* that entity is a matter of *having* a certain function. Whatever performs that function thereby realizes the entity of which that function is characteristic. We hasten to add that the critique of multiple realization we develop in the following chapters does not depend on the correctness of our theory of realization—perhaps some other theory better captures the characteristics of realization we outlined above. Rather, we shall argue that the relationship that many psychological state kinds bear to brain state kinds is compatible with their identification, and therefore we do not need to invoke the hypothesized realization relation for those cases.

These days, speaking of realization in terms of properties and property instances is common. Thus far, we have spoken variously—some would say sloppily—of the realization of entities, properties, states, kinds, objects, etc. In our defense, the conception of realization we favor does not, *prima facie* and in itself, put any limitation on which items may be the relata of realization. In our view, the relata of realization relations are whatever things can have and perform functions, in a broad sense. Below we will make the case for some relata rather than others being of the right sort to address questions about the metaphysics of mind. But that preference arises from our observations about the kinds of functions that psychological states possess, not from our account of realization itself.

Some philosophers worry that it is a mistake to understand realization in terms of functions because they regard that conception as too narrow—there are realization theories, they say, that are not functionalist theories. But this concern is itself rooted in a narrow way of thinking about what it is to have a function. We understand the notion of function quite broadly, so that any number of criteria or conditions might suffice for having a function c (cf. Melnyk 2003). In particular, we do not limit those conditions to just the mathematical or computational. Nor only to the causal. They might be semantic, teleological, epistemic, or whatnot; and they might be combinations of different sorts of conditions (Polger 2004a, 2007a). In contemporary discussions of the mind, the idea that psychological states are computational states of a probabilistic automaton, per Putnam's original hypothesis, has long since been replaced by the idea that psychological states have a definitive causal profile, which is to say, function.

Of course, you might say, for everything that there is, being that thing is just a matter of meeting some condition or other. If everything has identity conditions then everything has a "function." So what makes realization distinctive? That is, granting our expansive notion of function, the concern may flip from worries about the narrowness of the account to worries that it is too broad. Isn't it just trivial that psychological states are realized, given our inclusive notion of realization? We think not. According to advocates of realization theories, as we understand them, we must distinguish between the "first-order" properties of a thing in virtue of which it has the capacity to perform a function, and the additional "second-order" property of being something that has such first-order properties. Realizers must possess some "first-order" properties—that is, meet some conditions—in order to be realizers. However, the property that they realize is identified not with the "first-order" properties of the realizer (or realizers), but rather with the so-called "second-order" property of being such that it has some realizer or other. This point is crucial for the possibility of multiple realization, because it is the distinction between the "second-order" and "first-order" properties that allows for the possibility that sameness of "second-order" properties is accompanied by differences in "first-order" properties, i.e., multiple realization.

Despite appearances, the distinction between "first-order" and "second-order" properties is not a genuine distinction between two different kinds of properties.[5] Rather, it is between different ways of referring to those properties or kinds, allowing for the possibility that the same property might be characterized as second order on one occasion and first order on another. In the usual way of thinking, biological kinds are "second-order" kinds that are realized by the "first-order" kinds of chemistry. But those "first-order" chemical kinds might be "second-order" kinds from the point of view of the science of the realizers of chemical kinds, presumably physics. The idea is that the sciences that explain realized kinds use predicates and kind terms that are defined without use of the specific predicates and kind terms of the science of the realizers, but that pick out those things indirectly by reference to their relationships to one another as well as to their inputs and outputs.[6] That is, the "second-order" or realized predicates and kind terms are specified in terms of conditions (i.e. functions) that generalize or quantify over the "first-order" predicates and kinds of the realizer science. The idea is that purely functional kinds can be characterized without any mention at all of their realizers—entirely in terms of what the realizers do (i.e. their function, the conditions that they satisfy) rather than in terms of any details that identify particular realizers.

We like to illustrate the difference between realization theories and others with examples of familiar artifacts like mouse traps and cork-screws. There are many ways to build a mousetrap or a corkscrew. Corkscrews are devices for removing corks from standard retail wine bottles and similar corked bottles. Making something a corkscrew is that it does this job over a normal range of such bottles, when properly operated by a normal range of human beings. Further, let us stipulate,

---

[5] Funkhouser (2014) makes the same point, and takes this to be an objection to the functional theory of realization. But we think it just calls for some caution in stating the view. Here what is important for us is that advocates of the realization theory mean to draw some such distinction.

[6] Similarly, we recognize that there are no disjunctive properties, only disjunctive or non-disjunctive predicates by which we refer to various properties (Block 1997, Fodor 1997, Antony 1999, 2003, 2008). See also Jaworksi (2002) and Walter (2006).

in order to count as a corkscrew it must do this job by applying opposing forces to the lip of the bottle and the cork (or plastic plug, as the case may be), and that it must apply the force to the cork or plug by attaching to it by means of a screw that has been driven into the cork or plug. More could be said—the bottle should remain intact, the cork should not be deposited into the bottle, and so forth—but this is a good start on saying what it is to be a corkscrew.

Corkscrew is not a purely functional kind, because some features of its first-order realizers are specified. To wit, they must include a screw. And we assume that *screw* is a type in the realizing mechanical science, not only (or even) in the science of the corkscrews. But aside from the presence of a screw, the job description for corkscrews doesn't tell us which other components must be present, or how they must interact with one another. In particular, it does not tell us how the device should apply opposing forces to the lip and the cork, or how the operator should transmit energy to the device. It could be battery operated, steam operated, and so on. More familiarly, the device could involve a simple screw with a handle, on which the operator pulls; it could involve a single lever, as in the case of a "waiter's" corkscrew; it could involve two handles that operate pinions to withdraw the screw, as with a "double lever" corkscrew. The options are many. Being a corkscrew involves being something that does a certain job—that is, satisfies certain criteria or has a certain functional property. Perhaps each object that is a corkscrew satisfies these criteria in a different way. They may each have different realizer properties that allow them to extract corks within the constraints that characterize corkscrews. By having these realizer properties that perform the corkscrew-characteristic function, they each belong to the "second-order" or functional kind of things that perform the corkscrew function—and thereby realize the property or kind, *corkscrew*.

As long as we can draw the distinction between realizing and functional kinds, we can draw a structural distinction between realization theories and other theories. Using the realization theorist's vocabulary we can say that, according to the identity theory, important mental state kinds are identical to first-order kinds of brain states, kinds that are characterized with the proprietary resources of the neurosciences. In contrast, according to the realization theory, mental state kinds cannot be identified with first-order brain state kinds; that is, the members of those kinds do not belong to common first-order kinds of brain states. These are genuine alternatives, and so demonstrate that our account of realization is not so broad as to render trivial the question of whether any particular kinds, e.g., psychological kinds, are realized. They might not be. With this distinction we can clearly state the identity theory that we favor: Explanatorily important mental process kinds can be identified with brain process kinds, that is, brain process kinds that can be fully characterized using the resources of the neurosciences. This is how we understand the slogans of the classical identity theory, such as that "sensations are brain processes."[7]

Notice that questions about realization display a distinctive form. The appropriate question is not, "Is *corkscrew* realized?"—or, as we shall soon explore, "Is *corkscrew* multiply realized?"—but rather: "Is *corkscrew* realized—or multiply realized—by mechanical artifacts?" That question might, at least potentially, get a different answer than the questions, "Is *corkscrew* realized by fundamental particles?" or "Is the shape of a corkscrew realized by the arrangement of fundamental particles that occupies the same space as the corkscrew?" These are different questions. We are not interested in whether corkscrews or psychological states are realized *simpliciter*, or multiply realized *simpliciter*. In fact, we doubt that *realization simpliciter* and *multiple realization simpliciter* are useful notions. We think that questions about realization and multiple realization are always *specific* and *contrastive*. Is *x* realized by *y*? Is *x* multiply realized by *y*s and *z*s? In particular, we want to know whether psychological processes are realized or multiply realized by brain processes (or non-brain processes). Questions about realization and multiple realization are *specific* because they always ask about specified realized and would-be realizing kinds.

What determines which *x*s, *y*s, and *z*s are relevant to questions of realization and multiple realization? According to our view, questions about realization and multiple realization are always questions about the match or mismatch between the taxonomies of specific actual sciences

---

[7] What are brain process kinds? This is fundamentally a question for neuroscientists, and we fully expect that different neurosciences will study and articulate different process kinds in the brain. We will not be offering any unified account of brain process typing, but we will demonstrate by example that important cognitive and psychological kinds are identified with brain process kinds by neuroscientists.

and the entities and kinds within their respective domains.[8] We can ask whether genetic kinds are realized, or multiply realized, with respect to molecular biological kinds, as Kitcher does (1984). We can ask whether cellular neuroscientific kinds are realized, or multiply realized, by molecular chemical neuroscientific kinds—a question that Aizawa (2007) and Bickle (2003, 2006) have disputed.[9] And we can ask whether psychological kinds are realized, and whether they are multiply realized, with respect to neuroscientific kinds—the primary topic of this book. These questions are not only specific but *contrastive*: Answering each question requires us to contrast two (or more) scientific taxonomies to see whether they are in alignment or not.

As we have mentioned, it is common to speak of realization as a relation between properties or property instances. Our functional account of realization is compatible with the properties formulation but does not dictate that realization must be a relation between properties or property instances.[10] We have spoken of realization and multiple realization as relating kinds. For us, a functional kind dictates the nature of the relata of the realization relation. Some functions might be performed by properties or property instances, some by objects, some by processes, and so on. We frequently use the notions of property, process, and state generically, to discuss whatever entities might be the relata of the realization relation; and we suspect that in many cases the apparent variance is notational—it is an expository convenience to speak variously of the realization of states, processes, properties, etc. But it does seem to us that some cases are naturally described in one way rather than another. For example, when discussing corkscrews, we are asking about whether different kinds of concrete things, that is, objects, can be members of the same kind, viz., *corkscrew*.[11] In the contemporary literature about realization and multiple realization one typically thinks of the relevant functions as being characterized causally. So it is natural to say

---

[8] We take no stand on whether explanation is itself always contrastive (cf. van Fraassen 1980). We only claim that questions about multiple realization are questions about similarity or dissimilarity, and are therefore contrastive.

[9] See also Pauen (2002).

[10] We hope to avoid questions about multiple realization that arise from concerns about the nature of properties themselves. Others take multiple realizability to be intimately related to issues in fundamental ontology (e.g. Heil 1999).

[11] By objects here we mean not bare particulars, but ordinary propertied or "thick" objects.

that the realizers are the things that can have causes and effects, and those things are often taken to be causal properties or their instances. But we find it quite awkward to speak about the properties of a particular object, or of its parts, as realizing the property *corkscrewness*. We know what corkscrews do—what their function is; so we know what some thing must do in order to count as a corkscrew. We are less clear on what *corkscrewness* does, such that some other property or properties could do the *corkscrewness* job. Perhaps one might think about the causal powers that corkscrews have as giving the conditions for having the property *corkscrewness*; and then we could consider how the different properties of a certain object or its parts can contribute the powers constitutive of *corkscrewness* to the object (cf. Gillett 2003). But it seems more convenient to us to speak about particular objects being, or not being, corkscrews; that is, falling under the kind *corkscrew*. This looks like a case of objects realizing kinds.[12]

The crucial element to our view of realization is having a function, in our broad sense. We also claimed that, once we become specific about the entities or kinds we are comparing, the question of realization is not trivial—our account is not so broad as to imply that all properties, entities, etc. are realized. The question: "Is functional property x realized by base property y?" is not trivial—for x could be realized by y, or some other base property z. And, in particular, the question, "Is macroscopic entity x realized by microscopic or micro-based entity y?" is not trivial. Here, the question is not trivial both because x might be realized by a microscopic or micro-based entity z rather than y, and also because x's relation to y (or z) might not be realization at all. Instead x could be composed of, identical to, or constituted by y, to give a few examples. That is, the dependence relation between macro y and micro x might not be realization at all.[13]

---

[12] We don't know whether to reify kinds as properties—and thereby to reify functional kinds as properties picked out by "second-order" predicates. Complicating matters is that the "science of corkscrews" is plainly a fiction we have introduced to simplify our discussion, and matters quickly get complicated when comparing the taxonomies of actual sciences.

[13] Sometimes it is convenient to talk about an entity as a realizer even while raising the question of whether that entity in fact realizes some other, or is rather identical to or part of the other. Usually context will make it clear that when we use "realizer" in this non-committal way, but we will sometimes remind the reader by making it explicit that strictly speaking we are discussing a "would-be" realizer—something that would be a realizer if realization turns out to be the dependence relation in the case at hand.

2 Explanatory Realization Relations

For this reason, as we mentioned earlier, a successful account of realization must provide the basis for discriminating between realization and other ontological dependence relations. The fact that a certain item is a corkscrew is, plausibly, a fact about realization. If the story we told above is roughly correct, then being a corkscrew is a matter of performing a certain function. The *corkscrewness* of a corkscrew, if there is such a property, is realized. Other properties of the corkscrew, for instance its mass, are not realized, for having mass is not a feature of corkscrew functionality. Rather, corkscrews have mass simply because they are composed of matter. Because of the centrality of function to our account of realization, we reject accounts like Gillett's, according to which an instance of a property of a diamond, e.g. its hardness, is realized by instances of properties of the carbon atoms of which the diamond is composed (Gillett 2002, 2003; see Polger and Shapiro 2008, Shapiro and Polger 2012), for properties of diamonds are not functional kinds. Moreover, on the assumption that every thing has properties, and that every thing that exists (except for fundamental particles) is composed of something, Gillett must insist that every instantiated property is realized by instantiations of other properties—a claim that we reject.[14]

An account of realization should discriminate between realization and other dependence relations—other ways that things can be made up.[15]

---

[14] From our point of view, Gillett's (2002, 2003) account lumps together several distinct dependence relations. If an account of realization is not discriminate, then learning that x is realized by y would be uninformative. It would just tell us what we already knew—viz., that x depends on something else. This would be informative only in the minimal sense that it would eliminate the case in which x is fundamental, not depending on anything at all. It would be wrong to say that question never arises; sometimes it does. But in the case of the special sciences, we are concerned not with whether a thing is fundamental; we know that it is not. Instead, we are concerned to discover and understand its dependency on other things. We suspect that corkscrews, diamonds, and piles depend on other things in different ways; and there Gillett's account is indiscriminate. For discussion of various approaches to realization with respect to multiple realization and reduction, see Kim (2009, 2011), Haug (2010), Morris (2010), and Endicott (2010, 2012).

[15] We don't claim to have made these distinctions precise; nor shall we attempt to do so herein. We take it for granted that metaphysicians distinguish different sorts of dependence relations. The understanding of each has a long history, as does the question of whether they can be unified into one fundamental dependence relation. More importantly, we take it that scientists—though, of course, not in so many words—are also concerned with the different kinds of dependence relations in the world. They distinguish between things that are made up by purely aggregative relations, and those that are only made up when

We call accounts of realization that have this feature "explanatory" accounts, because discovery that some thing is realized can help to explain the sorts of manipulations and changes it can survive. Moreover, and again in contrast to Gillett, it is informative because it does not posit realization everywhere. Some things are realized, and some are not. A pile of sand is an aggregate or mass. Aggregates exist only if their parts exist—those very grains of sand; but aggregates can survive the rearrangement of their parts. So a pile of sand depends for its existence on the continued existence of specific parts, but not upon any specific arrangement of them. Perhaps they have to be in contact or proximity, but that's about all. In contrast, a diamond is not a mere aggregate. Diamonds are macromolecules with a distinctive micro-organization: Their parts have to be arranged in a lattice structure. But neither aggregates nor composites are realized because they are individuated by the first-order properties of their constituents. Corkscrews, on the other hand, differ from both piles of sand and diamonds. Corkscrews, on our view, are realized entities. The realizers of corkscrews don't have to be of any particular sort—indeed, corkscrews can survive the replacement of parts of their realizers, whereas aggregates cannot survive the replacement of their parts. And the parts of the realizers of corkscrews don't require any specific organization. There is only one kind of organization that makes something a diamond; but there are many kinds of organization that can make something a corkscrew.[16]

---

components are organized in some way. We can mark this distinction by saying that some things are unorganized aggregates and some are organized mechanisms (Wimsatt 1976, Craver 2001, 2002). It would be nice to have a general account of what distinguishes aggregates, mechanisms, and so on. But for present purposes the important claim is only that some such distinctions are made.

[16] We don't know whether it is universally true of realized kinds that their realizers can be organized in various ways, but it is certainly an explanatorily and practically important feature of many realized kinds. Could there be something that is realized but has only one possible kind of realizer? Nomologically unique realizers could be common, for all we know. The actual but contingent history of the universe is a powerful constraint on the viability of realizers. Certainly if we consider what is possible given the actual course of evolution, we will expect to find evolutionarily unique realizers. But whether there might be something for which there is only one metaphysically or logically possible realizer is less clear. Some examples might be (Keaton and Polger 2014): if biblical monotheism is true, perhaps god is the modally unique realizer of the kind *the perfect being*. Or one could argue that 2 is the modally unique realizer of *the smallest prime number*. Both of these examples achieve the uniqueness by invoking superlative kinds. Even if they hold up, upon inspection, these examples suggest very little about realization relations in philosophy of science and psychology.

The question being posed when we consider the realization or multiple realization of psychological states by neuroscientific states is a discriminate one. That the kinds of psychological states taxonomized by broadly cognitive and psychological theories are realized by the kinds of brain states as taxonomized by broadly neuroscientific theories is not trivial. And the question we as metaphysicians of science ask, rather than as practicing scientists, is not whether psychological states depend on brain states. We already know that they do. The question is not even whether, for a particular psychological state, it depends on a particular brain state in human beings, or chimpanzees, or octopuses. Rather, as philosophers we ask: Given the best available scientific explanations in the mind sciences and in the brain sciences, is the best hypothesis that psychological state kinds depend on brain state kinds by being realized by them?

In order to answer the metaphysics of science question about psychological kinds and brain kinds, we need to assess the available evidence provided by the empirical sciences of minds and brains. This brings us back to the question of the relationship between realization and multiple realization. In this section we have sketched our view of the constraints on a theory of realization, and provided some qualifications that distinguish our approach from others. There are three important messages to remember.

First, realization is a particular variety of metaphysical dependence relation. In order for the discovery of realization to be explanatory and informative, the relation should be discriminate. Not all properties or entities are realized.

Second, questions about realization are always specific and contrastive. They always ask whether an entity or kind in one scientific taxonomy is realized by an entity or kind in a second scientific taxonomy. Answering the questions requires contrasting taxonomic categories. In our toy example, we imagine that there is a science of corkscrews and a science of mechanical artifacts, and we consider the best model of the relationship between those taxonomies.

Third, as illustrated in the story about corkscrew science, the claim that something is realized is an outcome of the inquiry, not a premise in it or evidence for it. When we examine the taxonomies of two sciences, we may find that they align closely, not at all, or in various complicated ways. If dualism were correct, then the taxonomy of the science of the mind needn't stand in any interesting relation to the taxonomy of brain science. If the mind-brain identity theory is correct, then the relationship between those taxonomies would be intimate, compatible with identification of some or all of the kinds they categorize. But if multiple realization arguments are successful, then the taxonomy of the mind sciences is incompatible with the identification of psychological entities or kinds with neuroscientific entities or kinds. If there is a systematic relation between minds and brains but that relation is not identity then we are in a position to wonder what the relation could be. Putnam's specific proposal was to suggest that the mind is a probabilistic automaton that is implemented by the brain. More generally, we can say that realization is a proposed mind-brain relation that is not identity, yet is strong enough to secure the physical legitimacy of psychological states, and is compatible with the extent of the generality of psychological kinds—with multiple realizability.

If this is right then we must evaluate the evidence for realization theories by first determining the extent of multiple realization. How neuroscientifically diverse are the bearers of psychological states? In order for some advantage to accrue to the realization theory, the generality of psychological states must be more than the mind-brain identity theory can accommodate, and the realization theory must maintain all of the other virtues that the identity theory satisfied. If psychological states are "multiply realized" but that fact does not allow realization theory to "score" all the points tallied in Table 1.1 from Chapter 1, then it won't matter that psychological states are "multiply realized" or "multiply realizable" in that way. We put the term "multiply realized" in scare quotes here to distinguish between the kinds of variation in the world that would be a problem for an identity theory and also provide evidence for the realization theory, and those that do not. Perhaps this is just a terminological choice. We could say, "Multiple realization is prevalent, but it is not even *prima facie* problematic for the identity theory. Multiple realization per se has no tendency to support the realization theory." But we prefer to reserve the idea of multiple realization (and the term) for just those kinds of variation that are at least *prima facie* incompatible with the identity theory and that provide at least *prima facie* evidence for the realization theory. We think there is much less of that sort of multiple realization than is widely supposed, and in the coming chapters we will explain why.

3 How Much Multiple Realization Is Enough?

Before proceeding we would like to draw attention to one more question about the dispute between identity theorists and realization theorists. The question concerns just how much variation must exist in the world in order to provide evidence against the identity theory.

When we talk about functionalist and realization theories, we have in mind pure versions of those theories—functionalist theories that identify psychological kinds entirely in terms of their functional or role properties. In the above discussion about an imaginary corkscrew science, we claimed that a device has to latch onto a cork or plug by means of a screw in order to count as a corkscrew. Consequently, all realizers of corkscrews have at least one feature in common—a screw. *Corkscrew* is not therefore a purely functional kind, but a "mixed" or "anchored" functional kind (Rey 1997, Polger 2004a). We will insist upon the distinction between pure and "anchored" realization theories. We think that anchored functional kinds are common, but also that evidence for such kinds is not evidence against the identity theory. Anchored kinds presuppose some important inter-taxonomic identities. Those are the anchors. So it is important to distinguish anchored from non-anchored realization hypotheses, because the former do not rule out a substantial explanatory value for identities. If the best overall model of psychology and neuroscience includes many anchored kinds, then this would count in favor of an identity theory on our view, because they presuppose some identities rather than being incompatible with them. Identities will have proven to play an important role in explaining, in this case, how a corkscrew performs its cork-removing role.

With that qualification, we come to the question of just how to score which theory is more successful. How much variation—how much multiple realization—is needed to vindicate the realization theory? Sometimes the disagreement between identity theorists and their critics is presented in a way that suggests that if even one psychological state cannot be identified with any brain state, then the mind-brain identity theory is refuted. That is, the mind-brain identity theorist is portrayed as making the universal claim that every psychological state kind can be identified with a neuroscientific state kind. Putnam suggests that this is the correct standard, writing:

the brain-state theorist is not just saying that *pain* is a brain state; he is, of course, concerned to maintain that *every* psychological state is a brain state. Thus if we can find even one psychological predicate which can be clearly applied to both a mammal and an octopus (say 'hungry'), but whose physical-chemical 'correlate' is different in the two cases, the brain state theory has collapsed. It seems to me overwhelmingly probable that we can do this. (1967 in 1975: 436–7)

We think this presentation of the burden of proof on identity theory rests on an uncharitable interpretation of both historical and contemporary identity theorists (Shapiro and Polger 2012). We do not intend to argue that each and every psychological state kind can be identified with a single and unique neuroscientific state kind.

What about the reverse construal—that the realization theorist must maintain that every psychological state kind is realized, and no psychological state kind can be identified with any neuroscientific kind? This is, in our view, much closer to an accurate portrayal of the rhetoric of early functionalists of the 1970s, 1980s, and 1990s—at least in their polemical moments. Putnam, for example, says, "We could be made of Swiss cheese and it wouldn't matter" (1973 in 1975: 291). That is, functionalism is the view that no psychological states are anchored in any brain states—hence, what we are made of is entirely irrelevant to psychology. Likewise, Jerry Fodor says, "If the mind happens in space at all, it happens somewhere north of the neck. What exactly turns on knowing how far north?" (Fodor 1999).

But in our view his is also an unfortunate way of framing the disagreement. We know that realization theorists and functionalists made such claims, but we think that they should not have done so. Fodor's quip suggests that neuroscience is irrelevant for psychology—psychology is functional, and mind-brain identities add nothing to our understanding. In response, Fodor's critics have sometimes offered up examples wherein neuroscience appears relevant to psychology. But this looks like a boring and aimless debate. The question of choosing between realization theories and identity theories should not be about counting up examples. The question is not merely whether there is any multiple realization, or some, or a great deal. The amount of multiple realization in the world is important, but counting cases will not solve the dispute.

Rather, we believe, the better question is something more like: Does the best overall model of psychological and neuroscientific processes make substantial and important use of identities?

As you may anticipate, our answer is: Yes, the best overall model of psychological and neuroscientific processes makes substantial and important use of identities. In that model (or, as is likely, set of models) important psychological process kinds are identified with brain process kinds, and those identifications are explanatory. But at present we are focused not on our answer but on the process by which one should arrive at an answer. We cannot answer the question merely by counting the identities, nor merely by noting the variety of psychological beings or the extent of their neurological variations. The number and heterogeneity of variations is only one factor among many that will come into play.

From our current perspective, then, it seems that textbook "all-or-nothing" identity and functionalist theories are themselves idealized— caricatures, even. And some philosophical disputes over them are practically parodies of reasonable arguments that one might pursue. We intend to pursue one such reasonable argument. We argue that contemporary mind and brain sciences make substantial and important use of identities, for example by invoking anchored kinds. We support our conclusion by demonstrating that the best overall model of the psychological and brain sciences is one that includes substantial and important explanatory identifications of psychological and cognitive process kinds with neuroscientific process kinds. This is the substance of our identity theory. As we understand it, the opposing realization theory claims that identities rarely or never play an explanatory role in the cognitive and brain sciences. This is the crux of our disagreement.

What counts as substantial and important use of identities? Rather than setting out some quantitative or qualitative litmus test at the start, we will have to make the case that the use of identities in the mind and brain sciences is substantial and important. We're willing to accept that challenge.

To demonstrate that the best overall model of the psychological and brain sciences is one that includes substantial and important explanatory identifications of psychological and cognitive process kinds with neuroscientific process kinds, we have to show that such identifications are available. To do that, we have to show that multiple realization is not as obvious and ubiquitous as defenders of realization theories have supposed. That result is in fact more important than the mere defense of the identity theory of mind because its ramifications cannot be underestimated. For according to the realization theorist, the existence of and justification for the cognitive and psychological sciences—indeed, of all "special" or non-fundamental sciences—depends on the truth of multiple realization and the consequent impossibility of productively aligning the taxonomies of the different sciences. So by arguing that the cognitive and brain sciences make substantial and important use of identities, we are arguing against a general framework for understanding all of the special sciences and their subjects. We will return to address these consequences directly in Chapters 9 and 10. But the writing is on the wall: realization theories of mind are just the first special science domino to fall if we are right about multiple realization.

3

What Is Multiple Realization?

1 Multiple Realization, Variation, and the Autonomy of Psychology

In the previous chapters we made the case for the importance of multiple realization, historically and philosophically, in the apparent defeat of mind-brain identity theories, the rise of functionalist and realization theories, and the general defense of the legitimacy or "autonomy" of psychology. Multiple realization is absolutely central to a widely accepted model of the so-called special sciences—those sciences other than physics, including not only psychology but also biology, economics, and so forth. And the core idea has now made its way outside of philosophy of science altogether, inspiring "non-reductive" or "functional" theories in ethics, metaphysics, and epistemology, among others.

We also argued that, in order to deserve the significance that has been attributed to it, the phenomenon of multiple realization must be such that psychological states or processes are Real, Causally efficacious, and the subjects of Explanatory psychology. Those are all virtues that the identity theory delivers, so if the realization theory is preferable it should also preserve them or give us an overriding reason to give up on one or more of them. We don't know of any physicalists who willingly give up on any of those virtues. Some contemporary dualists reluctantly abandon Causal efficacy; but physicalists generally regard that as only making a bad theory worse.

And of course, if advocates of multiple realization hope to show the superiority of their view over the identity theory, they must demonstrate that they are better equipped to accommodate the Generality of psychology. That is, they will need to make the case that the neuroscientific diversity of psychological beings is so heterogeneous that not even a sophisticated identity theory can accommodate it. Establishing this incompatibility is, after all, the main job of multiple realization. As we saw in the previous chapters, the presumed fact of multiple realization illustrates not only the availability of realization theories, but also, given the supposed extent of variation among psychological systems, apparently mandates their adoption.

As a consequence of being multiply realized, psychological states are supposed to be ineligible for reductive explanation. For no single reduction to a particular realizer would match the potential Generality of a functionalist or realization theory—at best we can get local or species-specific reductions. For that reason, identity theories purportedly fail to accommodate the full generality of psychological kinds, and thereby fail to vindicate the legitimacy psychological sciences. For with no general reduction of pain or memory, the reasoning goes, identity theorists face two choices: Settle for more local explanations, abandoning hopes of Generality; or appeal to "Frankenstein kinds", assemblages of the reducible kinds, e.g. {*pain-in-humans*, or *pain-in-dogs*, or *pain-in-octopuses*, or etc.}. The latter case could be accused of sacrificing simplicity for the sake of Generality. So the identity theorist's proposal would be either less simple or less General than a functional or realization alternative; and it might still fail to explain what is common among psychological states of various kinds. Either way, functionalist or realization theories are taken to have the upper hand; and either way, the advantage stems from the fact of multiple realization. For this reason, we hold that any account of "multiple realization" that does not deliver these benefits is not an account of multiple realization of the sort relevant to the metaphysics of the mind sciences.

It is imperative to draw a distinction between multiple realization and other kinds of variation in nature. The distinction is important because variation in nature is the norm, top to bottom and as far as the eye can see. Our world is *Heraclitean*, replete with change and variety. To put things dramatically: All is flux. Some philosophers mistakenly think that the ubiquity of variability and change makes the fact of multiple realization self-evident. On some views, variation of any sort suffices for multiple realization. For example, Carl Gillett, sometimes in collaboration with Ken Aizawa, offers an analysis of multiple realization that would distinguish realizers on the basis of differences in their parts at any level of composition (2002, 2003, 2007; Aizawa and Gillett 2009a, 2009b). But on our view, this approach entails an undesirable profligacy of distinct realizations for every kind, and undermines the significance of realization within debates over the autonomy of the special sciences (Shapiro 2004b, Polger and Shapiro 2008, Polger 2010).[1] It would be odd indeed if the autonomy of psychology from neuroscience could be secured in virtue of tiny differences in potassium atoms. Moreover, we argued that realization theories are proposals for how to accommodate the fact of multiple realization; so we regard the proper understanding of multiple realization and the evidence for it as issues that must be resolved before one can offer an account of realization. To presuppose an account of realization when trying to understand multiple realization is to put the cart before the horse (Polger 2015). Finally, and most importantly, the sort of variation in composition that interests Gillett and Aizawa is, by their own account, no obstacle to reduction; so it is not suited to do the work that we have shown multiple realization is supposed to do (cf. Gillett 2007, 2013, Aizawa and Gillett 2009a, 2009b, 2011; see also Lyre 2009, Gozzano 2010, Soom 2012, and Theurer 2013).

Advocates of realization or functionalist theories who think that every variation between would-be realizers counts as evidence against reduction may be surprised to learn that identity theorists, from the very start, have recognized the ubiquity of variation in nature. J. J. C. Smart, whose "Sensations and Brain Processes" marked a seminal contribution in the development of the identity theory, wrote in a subsequent essay:

Compare topiary, making use of an analogy exploited by Quine in a different connection. In English country gardens the tops of box hedges are often cut in various shapes, for example peacock shapes. One might make generalizations about peacock shapes on box hedges, and one might say that all the imitation peacocks on a particular hedge have the same shape. However if we approach the two imitation peacocks and peer into them to note the precise shapes of the twigs that make them up we will find differences. Whether we say that two things are similar or not is a matter of abstractness of description. (Smart 2007)

So according to Smart, despite the numerous differences between each topiary peacock, there is something that all of the topiary peacocks have

---

[1] We have similar concerns about the "liberal" account of multiple realization proposed by Funkhouser (2007, 2014).

in common. The presence of variation in nature does not stop the various peacock-shaped hedges from being objectively similar, and that is all that is needed in order to group them into a kind. In at least this sense, an identity theorist appears capable of recognizing the legitimacy of "peacock kinds" over and above "twig kinds"—of a macrophysical kind in addition to a microphysical one.

Hilary Putnam, recall, supposed that establishing neuroscientific generalizations that cover all of the examples of psychological states—in humans, dogs, and octopuses, for example—could be done only by concocting a big disjunction of the individual generalizations for each creature type. But this approach, he said, we "do not have to take seriously" (1967 in 1975: 437). Despite Putnam's dismissiveness, it seems that another identity theorist, Herbert Feigl, had taken that exact possibility quite seriously nearly a decade before Putnam's comment. Responding to critics who thought that the identification of the mental and the physical could be possible only if mental and physical statements were synonymous, and therefore intertranslatable, Feigl wrote:

> we repudiate the *logical* translatability thesis not because of the possibility, definitely contemplated, of a one-many Ψ-Φ correspondence. One could always formulate such a correspondence with the help of a general equivalence between statements containing single Ψ-predicates on the one side and disjunctions of statements containing several and various Φ-predicates on the other. (1958: 391)

Apparently, *pace* Putnam, the possibility that the mind-brain relation might be, in some sense, one-many was "definitely contemplated" and yet Feigl recommended an identity solution nevertheless. Notice that if all variation in nature counts as multiple realization, then Feigl's view would be a non-starter. But we think that his view is an open possibility. Sciences, as we shall emphasize, may form their taxonomic kinds in many different ways. The world is full of variation, it is true; so there are many boundaries that might interest a scientist. But multiple realization is a special kind of variation between kinds. The reason that not all variation in nature counts as evidence of multiple realization is that not any sort of variation can perform the numerous jobs for which multiple realization has been recruited.[2] In this chapter and the next

---

[2] Funkhouser (2014) defends a similar job description for multiple realization, but we think our account does the job better.

we identify and elaborate the special features of the kinds of variation that make for multiple realization.

2 Putnam's Recipe and the Basic Recipe

To understand the phenomenon of multiple realization, a useful starting point is Putnam's introduction of the multiple realization argument against what he called the "brain state" theory, i.e., the identity theory. In a classic passage that we discussed in Chapter 1, Putnam writes:

Consider what the brain-state theorist has to do to make good his claims. He has to specify a physical-chemical state such that *any* organism (not just a mammal) is in pain if and only if (a) it possesses a brain of suitable physical-chemical structure; and (b) its brain is in that physical-chemical state. This means that the physical-chemical state in question must be a possible state of a mammalian brain, a reptilian brain, a mollusc's brain (octopuses are mollusca, and certainly feel pain), etc. At the same time, it must *not* be a possible (physically possible) state of the brain of any physically possible creature that cannot feel pain. Even if such a state can be found, it must be nomologically certain that it will also be a state of the brain of any extraterrestrial life that may be found that will be capable of feeling pain before we can even entertain the supposition that it may *be* pain. (1967 in 1975: 436)

Here we get the core idea of multiple realization: Mammals, reptiles, and mollusks reveal a psychological *sameness*—namely, they all feel pain or can be in the psychological state of pain—while at the same time exhibiting an obvious and profound *difference*. Putnam assumes that because the animals belong to distinct species, their brains will differ. This is why it seems unlikely to Putnam that we will ever discover "neurophysiological laws that are species-independent" and therefore it is correspondingly unlikely that the identity theory can be made consistent with the actual extent of multiple realization in nature.

From Putnam's original presentation we can extract a first pass at a recipe for multiple realization:

Putnam's Recipe: Multiple realization occurs if and only if two (or more) systems are psychologically the same and neurophysiologically different.[3]

---

[3] As we noted in earlier chapters, multiple realization has been defended in the context of many sciences. Putnam's focus was psychology.

Plainly we now need to clarify what makes for psychological sameness and what makes for neurophysiological difference. But first, let us remember that Putnam believes himself to be advancing an empirical hypothesis, and consider the evidence he had for his proposal.

On the former point, he confirms that his claim is meant as an empirical conjecture, but nevertheless judges it to be "overwhelmingly probable" (1967 in 1975: 437). Still, he never denies the possibility that the identity theory might someday be vindicated. Perhaps, for instance, pain in all species corresponds to a single kind of brain state. He muses:

> It is not altogether impossible that such a state will be found. Even though octopus and mammals are examples of parallel (rather than sequential) evolution, for example, virtually identical structures (physically speaking) have evolved in the eye of the octopus and in the eye of the mammal, notwithstanding the fact that this organ has evolved from different kinds of cells in the two cases. Thus it is at least possible that parallel evolution, all over the universe, might *always* lead to *one and the same* physical 'correlate' of pain. But this is certainly an ambitious hypothesis. (1967 in 1975: 436)

For readers accustomed to thinking of multiple realization as obvious, what is surprising in this passage is not the claim that the identity theory fails on empirical grounds. That we expected. Rather, the first surprise comes when Putnam immediately provides an example of commonality despite prima facie variation. Indeed his example is the camera eye, something whose complexity of organization even Charles Darwin admitted strikes us as unlikely to have ever evolved at all (1859: 186).[4] Yet, and as Putnam himself notes, the same kinds of eyes have in fact evolved independently more than once. Second, Putnam disavows an obvious objection to the claim that human eyes and octopus eyes are in fact the same, i.e. the fact that they are made of different parts organized in different ways. Human and octopus eyes are composed of distinct kinds of cells, have differently shaped lenses, and the optic nerve travels through the retina in human eyes but not in octopus eyes (Land and Fernald 1992, Fernald 2000). Yet according to Putnam, human eyes and octopus eyes are of the same kind, despite occurring in different species and despite being made of different parts and despite differing even in their functional organization.

---

[4] Darwin, of course, immediately goes on to explain why our difficulty imagining the evolution of eyes is no evidence against the fact of their evolution.

As it happens, we think that Putnam is right to draw this conclusion—and we shall have more to say about eyes. But it is surprising that this conclusion is Putnam's. By parity of reasoning we would expect him to say that human brains and octopus brains might well be alike. But he does not. Why not? Why do brains differ from eyes, as far as the probability of multiple realization goes?

Putnam says, "it is a truism that similarities in behavior of two systems are at least a reason to suspect similarities in the functional organization of the two systems, and a much *weaker* reason to suspect similarities in the actual physical details" (1967 in 1975: 437). We doubt this inference as a general principle, but we can understand the line of thought that attracted Putnam. Consider some behavior, B, that members of distinct species exhibit. There will be some probability that the functional organization responsible for B in one species is the same functional organization that produces B in another. But we might then ask about what Putnam calls the "physical details" of these similar functional organizations. Again, there will be some probability that the physical details across similar functional organizations is the same. Considered in the abstract, and assuming that there is no common ancestor that also exhibited the behavior, the probability pertaining to just similarity in functional organization will typically be higher than the probability pertaining to similar functional organization *and* similar physical details. So the identity theory, requiring identity in physical details, seems to be less probable than functionalism.[5]

The advocate of multiple realization may therefore think that the case of convergent evolution of camera eyes is the exception that proves the rule: The discovery of common eye structures in humans and octopuses is unlikely. Although we have made that discovery, doing so goes hardly any distance at all in increasing the probability that human beings and mollusks will share common brain structures. For, by the reasoning above, the mere fact that humans and octopuses have somewhat similar behaviors—if that is indeed true—provides only very weak support for the hypothesis that they are neurologically similar.

---

[5] And, indeed, because functionalists needn't commit to the claim that same behavior is even evidence for the same functional organization, identity theory appears *far* less probable than functionalism.

Perhaps Putnam thought that neuroscience had already discovered that human and octopus brains are not similar enough to host the same brain states. If so, he doesn't tell us what evidence he has in mind. But before long advocates of multiple realization were appealing to what is by now a favorite bit of evidence: Neural plasticity (e.g., Block and Fodor 1972). Plasticity is, in the first case, the ability of neurons—or, more pertinently, groups of neurons composing organized brain areas—to perform a variety of tasks. One and the same brain area can perform different tasks, or functions; and different brain areas can perform the same tasks, or functions. Plasticity seems to demonstrate that slightly or radically different groups of neurons, even within the same individual human being over time, can perform the same psychological capacities. If psychological capacities within an individual brain over time are multiply realized, then surely there is multiple realization across individuals, and even more surely across species.

Later we will examine the evidential force that examples of neural plasticity provide for multiple realization. For now, by considering Putnam's discussion and the introduction of the idea of neural plasticity, we can make the recipe for multiple realization slightly more precise. Sameness of psychological state is to be understood in terms of sameness of function. Two creatures are in the same psychological state if they have the same functional organization, on this view.

The second part of the recipe for multiple realization concerns neurophysiological difference. But Putnam says that difference in the cells that compose eyes does not suffice to make human eyes differ from octopus eyes. He seems to think the salient point is that human and octopus eyes are "virtually identical structures" (1967 in 1975: 436). He doesn't explain exactly what he means by this, but it's not hard to figure out. Both vertebrate eyes and octopus eyes are camera eyes with lenses that focus light on a concave retina that is located at the back of the eye (Figure 3.1).

Despite the fact that these organs differ both in composition and structure—the vertebrate retina is "inside out" and the octopus retina is not—Putnam takes them to be "virtually identical" structures. Why?[6] The answer, we submit, is that vertebrate eyes and octopus eyes operate by the same optical principles: They both use an iris to control the amount of

---

[6] An answer we don't consider is that Putnam was entirely ignorant about the nature of eyes. His discussion shows that he was not.

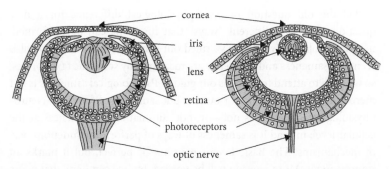

Figure 3.1 Cephalopod (left) and vertebrate (right) eyes. From Harris 1997: 2099. Copyright (1997) National Academy of Sciences, U.S.A.

light which is focused by a lens (and cornea, and the intervening medium) onto a photoreceptive surface. In particular, both kinds of eyes exploit the same optical regularities and do so using analogous structures. Vertebrate and octopus eyes are the same kind of mechanism—both in the ordinary sense of the term, as well as the technical sense recently prominent in the philosophy of science (Bechtel and Richardson 1993/2010; Glennan 1996; Machamer et al. 2000; Craver 2001, 2007; Skipper and Millstein 2005; Craver and Darden 2001, 2013; Bechtel 2008). Speaking colloquially, vertebrate and octopus eyes are the same kind of organ because they perform the same function (i.e., produce the same effects) in the same way (i.e. by exploiting the same optical regularities).

If two eyes are "physically speaking" alike when they do their jobs in the same way, then so too we should say that two brains are alike when they do their jobs in the same way. Or, better, two brain *areas* are alike when they do their jobs in the same way. This is what neurophysiological sameness must be, in Putnam's Recipe.

In short, the basic idea is that multiple realization occurs when the same psychological function is performed in different neural ways. And this idea can be generalized beyond the special case of the psychological and neural. In a slogan: Same but different. More precisely:

Basic Recipe: Multiple realization occurs if and only if two (or more) systems perform the same function in different ways.

Of course we've just traded the old question about psychological sameness for a new question about the sameness of function. And we've traded the old question of neurophysiological difference for a new question about the different "ways" that functions can be performed, neurophysiologically. But we've ruled out some alternatives. We've also seen that sameness and difference are measured by scientific taxonomies, so we need to attend to those as our guides. Psychological sameness is not mere sameness of behavior or output, but sameness of function. Neurophysiological sameness is not—or, not automatically—sameness at the cellular level; instead it is sameness of ways of performing functions, viz. of mechanisms. The Basic Recipe may not be perfect, but it marks an improvement. More revision will be needed, but we can hope that we've uncovered a foundation on which we can build a more tractable understanding of multiple realization.

3 Multiple Realization and Cognitivism

Before improving on the Basic Recipe, we should clarify more precisely the nature of the things that are supposed to be multiply realized. Multiple realization, we noted, amounts to more than just multiple causes for some effect. Belief in the multiple realization of the psychological begins with the cognitivist idea that some internal state or process produces (or is disposed to produce) an array of outputs given an array of inputs. The functionalist claims that a multitudinous variety of distinct internal states or processes can produce the same output from the same input. If two cognitive systems produce a certain behavioral effect but do not have anything *psychologically* in common, then the fact that they produce the effect in different ways will not be evidence for multiple realization. After all, we always knew that there was more than one way to produce an effect.

We emphasize this point because some advocates of multiple realization are, in our view, too quick to posit common psychological causes when they observe common effects. The result is that they think they find evidence of multiple realization where in fact there is no common psychological process to begin with. It is as though the search for multiple realization tempts us to fall back on behaviorist conceptions of psychological processes, according to which psychological processes are identified with stimulus-response patterns. But multiple realization is part of a cognitivist approach that is Realist about the cognitive states and processes that intervene between stimulus and response, between input and output. Multiple realization is a claim about those intermediate processes.

Consider recent research showing that honeybees can learn to identify individual human faces (Dyer et al. 2005; Avarguès-Weber et al. 2010).[7] Honeybees were trained to forage for a sucrose solution. In the experimental task, the sucrose solution was presented in front of photographs of human faces. The solution in front of the target face was enhanced with extra sucrose (reward), and the distractors were diluted with a salt solution (punishment). After training in this environment, the bees were tested in non-reward trials, including trials with similar distractors, and with inverted distractors and targets, i.e., the photographs were upside down. The results showed that bees can learn to discriminate individual human faces. For example, one bee was almost 94 percent accurate on the day of training and 80 percent accurate two days after training.[8] Performance diminished when the distractors were very similar to the target, as judged by humans; and performance was poor when the images were inverted. The authors conclude, "The findings indicate that it is possible for the visual and neural system of an animal to learn to reliably recognise a human face, even though the animal has no evolutionary history for the task" (Dyer et al. 2005: 4713).

Is this an example of multiple realization: Face recognition in humans and face recognition in bees? Face recognition with and without specialized neural structures? Dyer and colleagues don't think so. They write, "The results in this current study show that even bees are capable of recognising human faces and thus supports [sic] the view that the human brain may not need to have a visual area specific for the recognition of faces" (2005: 4713). They suggest that the fact that bees can recognize human faces without having any specialized and evolved brain areas for face recognition could cast doubt on whether human beings in fact make use of such a specialized brain area in face recognition. For bees have no neural structures that resemble the human fusiform gyrus—"fusiform face area." Rather than taking the bees' facial recognition ability to be evidence of multiple realization, Dyer goes in the opposite direction:

---

[7] We learned about this example from Ken Sufka.

[8] The number of bees that completed the tasks was quite low. In the second study, there were five bee subjects.

If facial recognition in bees can occur without a face recognition module, then human beings might recognize faces without such a neural module. Notice the underlying premise in this inference: Face recognition in bees and humans is likely to be realized in similar ways.

But the hypothesis of common mechanisms seems to have been premature. Critics counter that:

> Dyer's extrapolations about how recognition is achieved and whether or not it is facilitated by specialized brain regions are misleading. Face recognition is carried out by an automated and specific process in humans, which is known as *configural processing* (perceiving metrical relations between face features) . . . In their study, there is no clear evidence of configural processing and it is likely that the bees' recognition relied on specific features. (Pascalis et al. 2006, emphasis added)

This critique raises the concern we mentioned above: The bees may engage in a similar behavior to that of humans, but that by itself does not show that they are using a common cognitive process. There is no question of multiple realization of a cognitive capacity for face recognition if the bees are not recognizing faces by means of a cognitive or psychological process that they share with humans.[9] In this case, Pascalis and colleagues suggest that one marker of whether the bees and humans employ the same psychological process would be that they recognize faces by configural processing rather than by merely identifying specific features of target faces. Sameness of behavior is not sufficient for sameness of cognitive process.

Pascalis is plainly skeptical, leaving it open that we could discover that both human beings and bees recognize faces using configural processing. And indeed Dyer, Avarguès-Weber, and colleagues (2010) did just that: They performed a series of experiments designed to eliminate the possibility that bees succeed in the task merely by learning the specific features of individual faces. Success in the experiments required the bees to recognize configurations, just as humans do. They found that "Bees succeeded in categorizing face-like *versus* non-face-like stimuli using configural information and not only isolated features and low-

---

[9] Notice that, contrary to what some philosophers seem to expect, none of the scientists think the question of whether humans and bees recognize faces in the "same" way is settled by noting that human beings have camera eyes and bees have compound eyes, or that bees have a tiny fraction of the number of neurons that are in a human brain. Someone who made that proposal would be misunderstanding the issue that Dyer and Pascalis are disputing.

level cues such as the symmetry, center of gravity, visual angle, spatial frequency or background cues present in face-like stimuli" (2010: 600).

Although significant differences remain between the performance of human beings and bees in face recognition and configural processing, our focus here is on the nature of these processes. Is there an identifiable psychological state or process of "face recognition" or "configural processing" that both human beings and bees implement? There can be no multiple realization of a psychological state or process if there is no common psychological state or process present in both kinds of creatures.

Notice that "face recognition" in this context is a behavior, not a cognitive process that causes the behavior—that is why Pascalis et al. put emphasis on how the task is accomplished. "Configural processing" sounds like a psychological process. But it turns out that the kind of "configural processing" for which Avarguès-Weber et al. found evidence is defined only as "sensitivity to first-order relations" where sensitivity is understood in terms of differential behavior (Avarguès-Weber et al. 2010: 593). And they concede that although thinking of the bees as forming a kind of gestalt or feature-bound representation "constitutes an appealing framework to interpret the performance of the bees . . . so far the evidence obtained is contradictory and does not allow concluding that such a processing form is available in bees" (2010: 600). And "no evidence allows discussing Maurer and colleagues' third level of processing, 'sensitivity to second-order relationships', in which distances between features are perceived and used for discrimination" (2010: 600). The sort of "configural processing" that is observed is itself a pattern of behavior, not a cognitive mechanism that produces the behavior. According to Dyer, the missing evidence that bee face recognition might involve cognitive processes similar to those in mammals could include evidence of top-down processing, context effects, or flexible use of different strategies (Dyer 2012; cf. Menzel and Giurfa 2006).

So it seems as though the jury is out on whether the surprising ability of honeybees to recognize human faces is produced by anything like the psychological processes involved in human face recognition. The reason for uncertainty is that all of the researchers recognize that similar complex behaviors can be produced by creatures that do not have anything *psychologically* in common with one another. This fact leads some philosophers and scientists to be skeptical about whether non-human animals have any psychological capacities at all; but that is not our worry. Our point, rather, is that evidence for multiple realization has to begin with evidence of common psychological states and processes that are realized in different ways, not merely evidence of similar behaviors with various neural causes.

As philosophers, we leave it to the scientists to tell us whether bees and human beings undergo common psychological processes. It is not up to us. We cannot simply take the fact that human beings and bees exhibit face recognition behaviors and postulate that both human beings and bees have the property of *being face recognizers* or implement a process of *face recognition*. Whether there is a common psychological process that explains the behavior of human beings and bees is not a question of grammar, or of whether we can coin new predicates or nouns. It is a question of evidence and explanatory success. And in the case of honeybee face recognition, the science is incomplete. As Dyer himself says, "bees exhibit behaviors consistent with our current definition of cognition, but currently we know very little about the mechanisms underlying cognitive type behavior in bees" (2012: 394).

Multiple realization requires more than that there is a behavioral effect that has different causes in different creatures. Multiple realization requires that there be a common psychological state or process in different creatures—whether that common state causes the same behaviors or not—and that that common psychological state has different realizers in different creatures. There is no question of multiple realization unless there is a common psychological process to be multiply realized.

4   Varieties of Multiple Realizability

Claims of psychological multiple realization must start with the well-supported hypothesis that the creatures or systems under investigation are psychologically alike. According to the Basic Recipe, the systems must be the same with respect to their psychological function. Multiple realization requires "same but different" not just "kinda the same but different." "Kinda the same" opens the door to the possibility that variation in realizers does nothing more than explain the variation in psychological function—a point to which we will return at length. For this reason, multiple realization requires that the psychological functions be of exactly the same kind (Polger 2002, 2004b; Shapiro 2004, 2008).

When we think about the various differences in the world that might make for multiple realization, we consider two kinds of variation. On the one hand, there is the difference in the number of distinct would-be realizers. Are there only some, or many, or very many creatures that differ from us but that nevertheless have exactly the same psychological states? On the other hand, there is, roughly speaking, the magnitude of the differences between those creatures and the benchmark, viz., human beings. Do they differ only a little bit, very much, or so entirely as not to resemble us and our brains whatsoever? Both of these variables are important, because the claim that multiple realization is incompatible with identification and reduction is most plausible if the variation is both frequent and large. If the differences between realizers are few or small, this opens the door to useful psycho-neural identifications. So modest amounts and degrees of variation in realizers are not even prima facie problems for reductive and identity theories.

On the other hand, the greater the difference between would-be realizers and those present in human beings, the less confidence we generally have that those systems share exact similarities with human psychology. The idea that realizers that differ extensively from those that support human psychology can nevertheless support human-like psychological states is a powerful motivation for functionalist and realization theories, but the evidence for this idea is rather murky. Consider the view that every and any suitably organized system, regardless of its composition, can realize exactly the same psychological functions. That is, that the variation in the realization of psychological functions is both ubiquitous and massive. This view, which we call *Radical Multiple Realization* (Polger 2002, 2004b; Shapiro 2004), seems to be suggested by Putnam when he says, "We could be made of Swiss cheese and it wouldn't matter" (1973 in 1975: 291). How could it not matter if our "brains" were made of Swiss cheese? Of course it matters. It's terribly unlikely that Swiss cheese could ever be "suitably organized" to perform psychological functions. Bubbles aside, blocks of Swiss cheese are not organized at all; they are mere aggregates. As such, they violate what Ned Block calls the Disney Principle: "In Walt Disney movies, teacups think and talk, but in the real world, anything that can do those things needs more structure than a teacup" (1997: 120). So thinking cheeses are ruled out because they simply, as a matter of fact, cannot be suitably organized—they are not structured.

What is the point of thinking about copper, cheese, and souls, then? Radical Multiple Realization comes into play, in our experience, when advocates of multiple realization try to take seriously the empirical viability of their position and then start to worry about whether the evidence is on their side. Facing challenges about the empirical evidence for actual multiple realization, they retreat to the *possibility* of multiple realization— i.e., to multiple *realizability*. Perhaps the philosopher we imagine here is that notorious Strawperson who lurks in seminar rooms, or who only takes a position on multiple realization at hotel bars during philosophy conferences. But let us consider Strawperson's argument. As we see the matter, the best evidence for the multiple realizability of psychological states is their actual multiple realization. Suppose we have empirical evidence for the claim that systems of many physical compositions can have exactly the same psychological functions. We could have good inductive evidence for this claim if we observed many different systems that perform the same psychological functions in distinct ways. Even if we didn't observe anything close to indefinitely many systems doing the same function in different ways—and how could we?—we could observe enough diversity among the systems to infer that there are few if any constraints on the kinds of systems that can perform those psychological functions. For example, we infer, despite having observed only a small and finite sample of mousetraps, that there are indefinitely or infinitely many ways to be a mousetrap. We imagine that this is implied by our best theory of mousetraps. But not everything goes. Cheese is often used in mousetraps; but for obvious reasons it would be inadvisable to build a common household snap trap entirely out of cheese.[10] But this discussion of mousetraps seems to show that we can potentially have empirical evidence for multiple realizability.

But suppose that all evidence suggests that psychological functions come about in more or less the same ways in various creatures? Suppose, for example, that all the human-like psychological creatures we can find have neuronal systems more or less like our own. Such a finding would not eliminate the possibility of multiple realization, but it would be a

---

[10] We can nevertheless imagine mousetraps made entirely of cheese—perhaps involving a 200 kg wheel of gouda. The point, however, is that, "Even if there are many possible implementations of a given process, there may not be an infinite number of them" (Chatham and Badre forthcoming: 5).

tremendous setback for its defenders. Putnam advanced functionalism as an empirical hypothesis, on the grounds that it better fits the actual evidence as he saw it. Lacking evidence of actual multiple *realization*, what kind of evidence for multiple *realizability* could we have? We think that the right answer to our question is to admit that, in the absence of evidence of actual multiple realization, we should be very cautious indeed about insisting that mental states are nevertheless multiply realizable.[11]

This is where Strawperson usually steps into the debate. Strawperson doesn't think that we need to consider the evidence for multiple realizability. Strawperson doesn't care whether we have observed any actual cases of multiple realization, because what matters is only the *modal* claim that multiple realization is possible. Strawperson usually credits Jerry Fodor, or Jerry's Granny, with this idea. Indeed, Fodor anticipates our question, and moves to block the conclusion that lack of multiple realization is evidence against multiple realizability. He says, "The functionalist would not be disturbed if brain events turn out to be the only things with the functional properties that define mental states. Indeed, most functionalists fully expect it will turn out that way" (1981: 119). But can the functionalist really be so sanguine? The empirical result we're pondering is certainly compatible with a functionalist or realization theory. Yet if all the psychological systems we know about are in fact brain-based systems, what grounds the confidence that some need not be? Why think that multiple realizability is true for psychology if we don't have any examples of actual multiple realization?

The Fodorian is not without resources. One might still maintain that—like our imagined theory of mousetraps—our best overall model of cognitive systems makes it certain, or nearly certain, that psychological capacities are multiply realizable, even if we have no examples. That is, we could still have good theoretical reasons for accepting multiple realizability. That evidence would have to come from psychology and neuroscience. It would not be evidence of multiple realization, per se; but it would be evidence that supports a total model of psychological systems that permits multiple realization. For example, some philosophers and cognitive scientists seem to think that the fact of psychological multiple realizability follows from the success of computational

---

[11] This is not to say that we endorse as a general principle that absence of evidence is evidence of absence.

explanations in the cognitive sciences. We shall consider the actual evidence for multiple realization, including this kind of evidence, in the coming chapters.

But the truth is that we rarely encounter a defense of multiple realization that argues by means of theoretical models in this way. More frequently, we encounter reasoning that proceeds as follows:

> What I have been doubting is that there are neurological natural kinds co-extensive with psychological natural kinds. What seems increasingly clear is that, even if there is such a co-extension, it cannot be lawlike. For, it seems increasingly likely that there are nomologically possible systems (namely, automata) which satisfy natural kind predicates in psychology, and which satisfy no neurological predicates at all. Now, as Putnam has emphasized, if there are any such systems then there are probably vast numbers, since equivalent automata can be made out of practically anything. (Fodor 1974: 105)

It is worth walking through this line of reasoning carefully. Fodor begins with the idea that neurological kinds are not co-extensive with psychological kinds—that psychological kinds are in fact multiply realized. But then he retreats to the claim that even if they are in fact co-extensive—even if there is no actual multiple realization—that must just be a coincidence. Fodor asserts that any observed co-extension "cannot be lawlike." Why? What reason do we have for thinking, in the absence of actual evidence of multiple realization, that psychological states must be multiply realizable? (They "must" be multiply realizable if they "cannot" be nomologically or lawfully coextensive.) The reason, Fodor says, is that "it seems increasingly likely that there are nomologically possible systems (namely, automata) which satisfy natural kind predicates in psychology, and which satisfy no neurological predicates at all." That is, the reason that psychological states must be multiply realizable is that it is likely that multiple realization is possible. That, of course, is question-begging (Shapiro 2004).

Perhaps we should put the weight of this passage on the example that Fodor cites: The reason that psychological states must be multiply realizable is that it is "increasingly likely" that "there are" possible but non-actual systems (namely, automata—computing machines) that have psychological states. But this response is puzzling. Lacking evidence of actual multiple realization, what could make it "increasingly likely" that thinking automata will be actual? To be sure, if Fodor could point to an actual automaton that possessed human psychological states, and could assure us that the mechanisms of this automaton differed relevantly from a human brain, then he'd have made a strong case for multiple realization. Instead, Fodor seems to be enthused by the prospects for computational explanations in the cognitive sciences. Yet the justification for that enthusiasm is something for which we can demand evidence—and we will do so in Chapters 7 and 8. The lesson is that the Fodorian path to multiple realizability is not an alternative to providing evidence, though it may require evidence of a different sort.

Sometimes Strawperson looks to David Lewis' theory of the meanings of theoretical terms (Lewis 1970), rather than Fodor, to justify belief in the possibility of even radical forms of multiple realization. According to Lewis, theoretical terms—including, for example, mentalistic terms—can be defined by the role they play in a theory. "Pain" is defined as that state which plays a certain role—performs a certain function—in a psychological theory. And because the performers of functions can vary, just as different actors may play the role of Hamlet, we can be sure that the things to which "pain" refers can be various. Indeed, Lewis seems to hold that this account is perfectly general, applying at least to all theoretical terms except for those that appear in fundamental physics, which presumably do not describe functional roles (1994).

Our disagreements with the Lewisian are numerous and sometimes subtle, but two are worth mentioning. First, the Lewisian account of the meanings of theoretical terms presupposes a general account of meaning. If we adopt that account, then no explanatory use can be made of the distinction between kinds that exhibit multiple realizability and kinds that do not. For in the Lewisian picture, every non-fundamental kind is multiply realizable. The early identity theorists, if this is correct, were wrong simply because they adopted an incorrect theory of meaning. Now, we find Lewis' theory of meaning suspect and see no compelling reason to adopt it. But, more importantly for our present purposes, it is plain that this disagreement over the semantics of terms will not be resolved by evidence about multiple realization.[12] So, and this is the second point, for the Lewisian the question of multiple realization is

---

[12] We are not saying that the semantic dispute is immune from evidential support—on our view, it is not. Rather, the point is that evidence about brains and psychological function is of no special importance to evaluating that general semantic theory, whereas it does have a special importance for the multiple realization dispute that concerns us.

not an empirical one, in the sense that Lewis did not offer his version of functionalism as an empirical hypothesis about the nature of minds.

The literature about functionalism in philosophy of mind has often distinguished between the Putnamian "role" version of the theory and the Lewisian "realizer" version (see, e.g., Block 1978). But the literature tends to downplay the fact that Putnam and Lewis advanced theories of quite different sorts—one an empirical hypothesis, and the other not. Recognizing this, we can now clarify the dispute between ourselves and advocates of multiple realization. Our concern is with multiple realization as a claim about the nature of our world that *a fortiori* lends itself to empirical adjudication. In contrast, we are not concerned to defend any view of the semantics of terms, even psychologistic terms. And although we shall not defend the claim herein, we are inclined to think that the Lewisian must be able to reconstruct our dispute over the evidence for multiple realization within her framework, so that the empirical question of multiple realization is as legitimate for her as it us for us.[13] But, most critically, appealing to Lewisian considerations is not a viable fall-back strategy for someone who wishes to defend the Putnamian hypothesis. Because the latter stakes out an empirical claim, it does no good to look to an a priori theory of meaning when short on evidence.

Are we denying the possibility of aliens or robots that can think? No, we are not saying that. But one must be cautious. Presently we have not observed automata with psychological states (or, at any rate, psychological states anything like our own). Nor, as we shall argue in Chapters 7 and 8, do any specific considerations from neuroscience, evolutionary theory, or computer science suggest that the best overall theory of the nature and science of minds is a theoretical model according to which automata can have psychological states. For now, we are merely remarking on the unreasonableness of the demand that every account of the metaphysics of psychological states be compatible with psychological automata. The possibility of psychological automata is an implication of some theories, not a datum to be explained.

In light of this point, the error in a popular line of reasoning becomes clear. Consider this claim from Richard Fumerton:

---

[13] Just as Sydney Shoemaker never thought that his functionalist causal theory of properties settled the question of functionalism in philosophy of mind (Shoemaker 1980, 1981).

it is hardly the case that the philosopher needed to wait on empirical research to settle the relevant philosophical issues. We would have to be stupid not to realize that it *might* turn out that creatures with interestingly different kinds of brains could still have the same mental states that we have. We didn't need to consult cognitive scientists to reflect on the relevant *modal* question. All we needed to do was watch enough episodes of *Star Trek* to realize that we had better understand mental states in such a way that it is at least *possible* for the same mental state to have a radically different physical base. (2007: 60–1)

But this is preposterous. Empirical results certainly bear on the modal question. How the world could be depends in part on how the world is in fact. Science fiction is a useful prod to the imagination. But to put the point bluntly: Stories about Commander Data are not data.

Furthermore, advocates of multiple realizability assume that if robots and extraterrestrials could realize psychological properties then we can be confident that they do so in a way that differs relevantly from brain-based psychological creatures, and that we can know that fact without knowing anything about how they work. Yet what justifies that assumption? Our ignorance of the inner workings of possible but non-actual psychological beings can hardly give us confidence in the claim that they differ relevantly from us. Why should it? For if their psychological states are realized in the same way that they are realized in us, then the fact that they are composed of different stuff is not important—just as Putnam rejects the relevance of the difference in cells that compose human and octopus eyes. Ectoplasm and epithelial cells could both be lenses in the same way if they could both focus light by refraction.

We do not deny that *if* cognitive digital computers are possible, then multiple realization is probably true and the identity theory is probably false. The possibility of thinking machines favors a functionalist theory of minds. But none of this lends evidential support for the hypothesis that mental states are multiply realizable. It is odd, to say the least, to insist that the actual relationship between psychological kinds and neuroscientific kinds could be settled by reading science fiction. Nor do the ambitions of artificial intelligence decide the question of multiple realization, although they surely amount to a wager on the outcome.

To reiterate: The actual multiple realization of mental states was supposed to be evidence in favor of functionalism and realization theories. If we fail to discover such evidence, advocates of functionalism cannot fall back on the "mere" modal claim of multiple realizability—

the possibility that functionalism is true. For as we have gone to great lengths to explain, without some specific scientific and theoretical reasons for thinking that multiple realization is to be expected even in the absence of evidence, its defenders are left with only the question-begging assertion that functionalism is possible. And the possibility of functionalism is not evidence of the truth of functionalism.

Finally, even if we had a good reason to believe in the possibility of multiple realization, it is not obvious how that should weigh in the theoretical scorecard between identity theories and realization theories. Consider, for instance, the desire for a General psychological theory, i.e. one that captures generalizations over a variety of physically distinct cognizers. Should a psychological theory that recognizes the possibility of mentalistic Martians score higher for this reason than a theory that cannot accommodate this possibility? Why regard one theory as more General than another simply because it allows for non-actual possibilities that the other does not? The same science fiction stories that feature intelligent aliens and robots also feature creatures whose "biology" is based on methane, crystals, and metals. Should the possibility of such life forms, by itself, compel us to expand current biological theory in order to account for these merely possible life forms?[14]

On our view, multiple realization should be understood as something like the claim that systems of many (even indefinitely many) different kinds of physical structures can have exactly the same psychological functions. That kind of multiple realization would be enough, at least prima facie, to raise trouble for identity theories. We don't have to look for evidence that literally anything (e.g. cheese, cauliflower, toilet paper) could be part of a psychological system. By the same token, if we don't find evidence of actual multiple realization, then advocates of functionalist and realization theories cannot confidently retreat to the claim that psychological states are possibly multiply realized—that they are multiply realizable. Asserting the possibility of multiple realization is not an

---

[14] Brad Weslake argues that evaluations of a theory's explanatory "depth" or power ought sometimes at least to include its ability to accommodate nomologically impossible phenomena (2010). But we do not think he would endorse attributing a speculative feature to actual systems on the grounds that doing so would permit the formulation of a more powerful theory: The theory must fit the data, not our epistemic wishes. And of course depth is not the only theoretical virtue to be considered.

alternative to finding evidence for that possibility. Evidence of multiple *realizability* is just as necessary as evidence of actual multiple *realization*. That the world could be different than it is in fact—that psychological states are multiply realizable even if not multiply realized—is in itself a claim about the world.

4

Multiple Realization and Relevant Differences

1 Relevant Differences

The Basic Recipe for multiple realization is: Same psychological function, different neurophysiological way of performing the function. We are going to be looking for evidence that it occurs, or at least a good reason to expect it. We also argued that not every variation in a psychological system qualifies as multiple realization of a function because not every variation amounts to a different way of performing a function. Vision scientists regard human and octopus eyes as the same in type despite differences in their cellular composition. Cellular differences do not make a difference to how the eyes perform their function: They both focus light on a photosensitive surface in the same way. They rely on the same kind of mechanism.

Let us foreshadow some conclusions that we will defend in the coming chapters, and anticipate some questions. On our view, *eye* is a multiply realized kind, but not because of variation in cellular composition. That is not the right kind of variation. Human eyes and octopus eyes are the same kind of realizer of *eye*, namely, *camera eye*. But *camera eye* might be multiply realized: There are camera eyes with inside-out retinas (human) and camera eyes with outside-out retinas (octopus). And so on. The difference between inside-out and outside-out retinas may well count as multiple realization of retinas. But one cannot make that determination without examining the science of retinas. The differences between the cells in human and octopus eyes may well constitute a difference in the realization of part of an eye, or part of a part, etc. of an eye. But as it happens, it does not suffice for multiple realization of eyes themselves. And eyes do not inherit the multiple realization of their parts or realizers. Multiple realization doesn't percolate up in this way.[1]

How do we know which differences count as multiple realization and which do not? At their heart, these questions concern the kinds that figure in scientific explanations—or, in our earlier discussion of corkscrews, imaginary scientific explanations. When asking whether psychological kinds are multiply realized by neuroscientific kinds, we must attend to the taxonomies of explanatory kinds in which the cognitive sciences traffic. The sciences themselves set the terms of the investigation, with their own criteria of precision and specificity (cf. Couch 2009, Balari and Lorenzo 2015).

But furthermore, any investigation of multiple realization depends on certain idealizing assumptions. The investigation starts from the contrary to fact supposition that the relevant sciences are "finished," or at least not far from being so. Our conclusions are correspondingly fallible: If the sciences develop in an unanticipated way, then earlier answers to questions about multiple realization may change, for multiple realization is a fact about the world as our best sciences portray it.

This last point reinforces our conviction that the various sciences to which we must look to assess the possibility of multiple realization may well constrain each other in various ways. Accordingly, we shall have to consider carefully what sorts of autonomy they do or should have from one another. We cannot assume that the sciences of the mind have developed independently of one another. On the contrary, we know that they have developed in tandem, sharing methods, data, and more (Bechtel and Mundale 1999; Bechtel and McCauley 1999; McCauley and Bechtel 2001, Sullivan 2008, 2009; Bickle 2006, Churchland 1986). This co-evolution of the cognitive sciences is obvious to anyone who has

---

[1] The reason is simple: Multiple realization is a claim about the dissimilarity of two taxonomies, and dissimilarity is not transitive. P might be realized by Q, and Q might be multiply realized by R and S; but that does not imply that R and S are multiple realizers of P. One way this could happen is that the difference between R and S is relevant to kind Q and irrelevant to kind P.

studied psychology or neuroscience. But, as we will argue in Chapter 10, we needn't take such co-evolution to diminish the autonomy of individual sciences.

In the case of eyes, there are many different sciences that study vision, but only some for which the eye itself is an explanatory component. As it turns out, those sciences tend to differentiate eyes according to the various optical principles by which they direct light to a photosensitive surface, rather than by the various cellular kinds that compose them (Land and Fernald 1992). The important point is that not every difference makes for multiple realization. Only some variation in nature is relevant to considerations of multiple realization.

The science that deals with a potentially realized or multiply realized kind will dictate the boundaries of the kind category—and thereby tell us which differences in performance qualify as differences in function. Relevant differences are those that, according to that science, make for differences in kind. The sciences of eyes tell us which photosensitive organs count as eyes, and which do not. Human skin is light sensitive, but nevertheless not an eye. Human eyes and octopus eyes have the same function; they are both organs for forming images on photosensitive receptors. Human skin and human retinas have relevantly different functions, although they both show sensitivity to light. Likewise, the sciences that deal with the putative realizers tell us which differences in them amount to different ways of producing their effects and which do not. That is, they define those differences that suffice to distinguish one kind of realizer from another. As before, the relevant differences are those that make for differences in kind according to that science.

We therefore need to revise the Basic Recipe for multiple realization to recognize that only certain differences contribute to multiple realization. The slogan is not just "same but different," but rather "relevantly the same and relevantly different." This doesn't trip off the tongue as nicely as the Basic Recipe slogan, but it is a better theory:

> Revised Recipe: Multiple realization occurs if and only if two (or more) systems perform relevantly the same function in relevantly different ways.

According to this recipe, in order for two things to count as performing the same function they need not be exactly alike in every way, they only need be relevantly alike. Likewise, in order for them to count as performing that function in different ways, the differences between them must be relevant to how they perform that function (Shapiro 2000, 2004).

This appeal to relevance may seem too obvious to mention, but it in fact marks a significant advance beyond ordinary ways of conceiving multiple realization. Without reflecting on the issue of relevance, claims of multiple realization become trivial. If variation of any sort at all constitutes multiple realization, the thesis begins to sound more like an a priori commitment than an empirically risky conjecture. It hardly sticks its neck out, as one would hope a scientific hypothesis should do. This is why, as we mentioned earlier, we think that philosophers like Ken Aizawa and Carl Gillett (2011; Gillett 2003), who allow variation of any sort to distinguish between realizations—as little as a difference of a single molecule—are heading down the wrong path.

Attending to relevant differences makes sense. Eyes don't have to be exactly alike in every way—having all and only the same features. To be eyes they need to be alike only with respect to the features that make them eyes. Which differences are relevant to their sameness (or difference) of function is determined by the sciences that make use of the taxonomic kind *eye*: That explain their functioning, or that appeal to their operation in order to explain some other phenomenon. If we want to know whether some prima facie distinct organs that do the job of eyes are the same or different—whether they count as multiple realizations of eyes—then we have to consider the features by which they differ, and whether those features count as different ways of being eyes. From the point of view of biological anatomy and mechanics, it appears that human eyes and octopus eyes are the same kind of mechanism, they do the eye job in the same way: They are camera eyes. That is, both the science of eyes and the science of light-focusing mechanisms count human eyes and octopus eyes as being of the same kind. Although the mechanisms by which they perform the eye function differ in various ways, those differences, from the perspective of ocular sciences, are not relevant to the way they do the eye job. Therefore this is not a case of multiple realization.

We offer this conclusion with caution for we are working with the cartoon idealization according to which there is a singular science of eyes, and making claims about what "it" would say. Yet we do not doubt that some sciences for which *eye* is a kind do in fact distinguish between human and octopus eyes. So we should not be misunderstood as claiming that every science of eyes must classify them in one way or another. Rather, our point is twofold. Sciences deploy many methods for evaluating the sameness and difference between things; they group their subject matters according to a variety of similarities and differences, often taking into account considerations that may surprise outsiders. Consequently there is no shortcut around comparing and contrasting the taxonomies of actual sciences, or samples of them such as those involved in individual explanations or models (cf. Richardson 2008; Couch 2005, 2009). The question of multiple realization is a question about actual sciences, and it is always specific and contrastive. The question is not, "Are eyes multiply realized?" It is, "Is the kind *eye* in science A multiply realized by kinds K_1-K_n in science B?"

We made this point previously with respect to our favorite example, corkscrews. First, we imagine a science of corkscrews, which determines which things count as corkscrews and which do not. Simple corkscrews, waiter's corkscrews, double-lever corkscrews, duck-rabbit shaped corkscrews, and innumerably many other kinds of mechanisms count as corkscrews. But, we imagine the science of corkscrews on the model of the sciences of eyes. So we imagine that just as some light-sensitive cells are not eyes, so too the science of corkscrews would say that some cork-removing devices fail to be corkscrews. We think it would say that corkscrews work by screwing into the cork, and that they normally pull the cork out of the bottle. For this reason, a number of devices that open bottles of wine are not corkscrews—devices that slide between the cork and the inner neck of the bottle to withdraw the cork, devices that inject pressurized gas into the bottle to force out the cork, and devices that push the cork into the bottle.

In addition, when considering whether different corkscrews are multiple realizations of the kind *corkscrew*, we imagine a science of mechanical artifacts that differentiates among members of the folk kind *corkscrew* according to which mechanical principles explain their operation. A litmus test is whether the same mechanical explanation can explain the operation of two devices. Thus, from the perspective of this science, waiter's corkscrews differ from double-lever corkscrews because different mechanical principles and explanations apply to them. In the case of a waiter's corkscrew, opposing forces are applied to the bottle and the cork by means of a hinged lever. In the case of a double-lever corkscrew, opposing forces are applied to the bottle and the cork by fixing the device against the bottle and using two handles with pinions to apply force to a toothed rack. Because these two devices make use of different mechanical principles—one levers, one rack and pinions, and so forth—we conclude that they do the corkscrew job in different ways, as considered from the point of view of our imaginary science of mechanical artifacts. In contrast, if we consider two waiter's corkscrews that differ in the material of which they are made (e.g., aluminum, stainless steel, or wood) or in their color (e.g., silver, black, or pink), the imagined science of mechanical artifacts tells us that these differences are not relevant differences (Shapiro 2000). The reason is that the same explanation of how the device does its corkscrew job applies to all of them, regardless of material composition and color. In this case, the explanation involves a hinged lever.

Note that even if the differences in material composition, according to our idealized mechanical science, are not relevant to judgments of multiple realization of corkscrews, they nevertheless amount to real differences in the corkscrews. A stainless steel waiter's corkscrew is presumably stronger than an aluminum or wooden model. Perhaps it would succeed in removing some particularly resistant corks, while the others would break or bend. Or, perhaps it would allow too much force to be applied, tending to break the cork or the bottle. But these factors, within limits, do not matter to its classification as a corkscrew. If a device while functioning properly always broke the bottle or never removed a cork, then it would not be a corkscrew.[2] A device shaped like a waiter's corkscrew but made of cheese, paper, or peppermint sticks is not a different realization of corkscrew, if it is a corkscrew at all. Neither, according to the corkscrew science we have imagined, is an electric drill, despite the fact that a drill bit is a screw and can perhaps be used to remove corks from bottles in a pinch.

Moreover, though this discussion of corkscrews invites the charge that our judgments of multiple realization depend on mere stipulations about which properties are important (e.g., rigidity) and which not (e.g., color), we believe this to be an artifact of the idealized example. In the case of legitimate scientific kinds, such as eyes, the sciences provide principled means for distinguishing relevant from irrelevant differences. The anatomical differences between mammalian camera eyes and insect

---

[2] Although, on some etiological views of function, e.g. Millikan (1984, 1989), a terrible corkscrew, or one that never works, might still have the function to remove corks.

compound eyes are differences relevant to their ways of being eyes. They make a difference to how light is collected and focused in each eye. On the other hand, whether the optic nerve travels through the retina, as it does in the human eye, or attaches to the back of the retina, as it does in the octopus eye, has been judged by eye researchers not to be a kind-differentiating feature of eyes in some explanatory models of eyes (see also Couch 2005, 2009).

2 Beyond Relevant Differences: The Official Recipe

The compositional differences between wooden, aluminum, and steel waiter's corkscrews are not relevant differences to their ways of being corkscrews, nor of being waiter's corkscrews. But being a waiter's cork-screw is such a simple job. There is no definite criterion for just how much force the corkscrew must apply, how long it should take, who should be able to operate it, etc. So the differences between wooden, aluminum, and steel models do not make for a difference in kind.

One of the factors that makes something a waiter's corkscrew is, we're imagining, having a lever. Differences in material composition may have consequences for differences in rigidity of the lever. Do the differences in rigidity among levers count as multiple realizations of the kind *waiter's corkscrew*? No. Rigidity of the lever is relevant; it must be in a certain range. For example, a device with an insufficiently rigid lever could not perform the waiter's corkscrew task at all—such as the cheese "corkscrew" considered above. But the rigid lever (or the rigidity of the lever, if you prefer) does the same thing in all of them. This is not a case of the same function done in different ways; it is a case of the same function done the same way. Among those things that do perform the waiter's corkscrew function, differences in the rigidity of the lever do not make for different kinds of mechanisms. Instead, the differences in the rigidity of the lever correspond to individual differences among waiter's corkscrews.[3]

---

[3] Although the differently rigid levers do not make for different realizations of *waiter's corkscrew*, rigidity itself might be multiply realized. Metals are rigid in different ways than woods—the former are rigid because of their isotropic molecular organization; the latter are (much less!) rigid because of their anisotropic laminar cellular organization (Ugural and Fenster 2003). But multiple realization does not percolate up, so this does not entail that corkscrew (*corkscrewness*) is multiply realized by different rigidities.

The preceding considerations reveal that "relevantly the same and relevantly different" is not exactly the right slogan, after all. We need to ensure that whatever differences qualify to distinguish one kind of realizer from another do so in virtue of how they contribute to the sameness of the functions of the realizers. They must count as different paths to the same end, not just variations within a single path to the same end. Ruth Millikan construes multiple realization similarly, writing:

> Sometimes different mechanisms that accomplish the same [sic.] operate in accordance with different principles; other times they represent merely different embodiments of the same principles. Or we might say, sometimes looking more closely at the mechanism helps to explain how it works; sometimes it reveals only what stuff it is made of. It is only the former kind of difference that makes interesting "multiple realizability." (1999: 61–2)[4]

Likewise, on our account, some variations in realizers amount to nothing more than individual differences, of the Heraclitean sort we mentioned earlier. Variation is everywhere in nature, but multiple realization is not.

Taking these considerations into account, we can formulate our official recipe for multiple realization (after Shapiro 2008, Polger 2009a, Shapiro and Polger 2012):

Official Recipe:

(i)   As and Bs are of the same kind in model or taxonomic system S1.
(ii) As and Bs are of different kinds in model or taxonomic system S2.
(iii) The factors that lead the As and Bs to be differently classified by S2 must be among those that lead them to be commonly classified by S1.
(iv) The relevant S2-variation between As and Bs must be distinct from the S1 intra-kind variation between As and Bs.

Clauses (i) and (ii) capture the requirement that multiple realizers must be "same but different." We use clause (iii) to express the idea that multiple realization requires A and B to be not merely different, but to be "relevantly different"—to be different in ways that are relevant to their performing the same function. Winged and waiter's corkscrews differ in ways that contribute to their cork-removing capacities; camera and

---

[4] Funkhouser (2014: 104) takes us to be elaborating on these "passing comments" from Millikan; but in fact we had overlooked this passage until Funkhouser directed our attention to it.

compound eyes differ in ways that are relevant to their light-sensing capacities. Differently colored waiter's corkscrews, in contrast, are the same with respect to their cork-removing capacities—they are samely the same, not differently the same.

Clause (iv) captures the "differently the same" part of our account: For multiple realization, the differences among would-be realizers must be "other" than mere individual difference. These differences need not be large differences—sometimes small differences may contribute to multiple realizations; but the variation must not merely map onto individual differences.[5]

Clauses (iii) and (iv) are necessary if multiple realization is to be an obstacle to reduction, i.e. to the identification of realized kinds with their would-be realizers. If members of the would-be realizing kind differ only in ways that have nothing to do with how the realized kind is individuated, those irrelevant differences will not block an identification.[6] Chemistry provides an interesting example of type individuation that disregards individual differences. The periodic table identifies chemical elements by their atomic number—the number of protons in the nucleus of an atom of the element. For these purposes, atomic weight—which varies according to the number of neutrons as well as protons—is simply not relevant. Elements of the same atomic number but different atomic weights are isotopes. But it is plain that isotopes of gold are not different ways of having the atomic number 79 (i.e., being gold)—they are not multiple realizations of gold—because they all have their atomic number in the same way: Namely, by having 79 protons in their nuclei. The kind *chemical element having atomic number 79* and the kind *chemical element having 79 protons* are one and the same kinds.

Likewise, clause (iv) is necessary because if the differences among would-be realizers are recognized to be differences within a kind rather than between kinds, then kind identification is not blocked.[7] Examples

---

[5] It is useful to remember that the S1 and S2 taxonomic systems may operate at the same or different mereological levels. An anvil and a corkscrew may be classified similarly by the things-that-can-be-paperweights science and differently by the things-that-can-be-anchors science (same level) and differently by the things-that-react-with-acids science (lower level). We speculate that this fact misleads Carl Gillett into thinking that the realization relation is itself multi-level or "dimensioned" (2002, 2003).

[6] See also Francescotti (1997, 2014).

[7] As we understand the account proposed by Funkhouser (2014: 115), it is roughly equivalent to our criteria (i)–(iii). But Funkhouser interprets our talk of different "ways" as

involving quantitative differences are most straightforward. The wings of sparrows may vary in size. Presumably there are some upper and lower limits on their size, beyond which they would not function as wings. So the size of the wings is relevant to their *being* wings. On the other hand, the variations among wing sizes of sparrows typically fall within this range. Those are relevant differences in the wings. But those differences correspond to differences in the flight abilities of individual sparrows. So differences within a range of wing lengths are differences that explain difference, rather than differences that explain sameness. It is the sameness of the wings that explains the sameness in wing-related functions; and the differences in the wings explain only the differences in wing-related functions. A wing anatomist would recognize a variety of sparrow wings to be wings of the same kind, even though they differ in wing-relevant ways such as size and weight.

Consider different kinds of watches (cf. Shapiro 2008). For the moment let us limit our attention to analog watches, those that display the time by means of hands whose position indicates hours, minutes, and seconds. Perhaps watchmaking is not quite a science, but it is at least a very refined craft. Watchmakers distinguish many kinds of watch movements (i.e., mechanisms), but two are most common: Mechanical and quartz watches. Let us focus on two distinguishing features of watch movements: The power source, and something that regulates the rate of the movement. In mechanical and automatic movement watches, the power is stored as mechanical potential energy in a coiled mainspring (Figure 4.1). This energy is provided by winding the watch, either by turning the crown or also, in the case of automatic mechanical movements, by rotors that wind the spring when the watch is moved.[8] As the mainspring relaxes, the potential energy is transferred to the surrounding "barrel" to which it is attached, and converted into movement that is then conveyed to the gears that move the hands of the watch. But without some sort of regulation, the gears and hands would simply spin

---

indicating that we conflate realization with other determination relations, rather than understanding it as indicating our "differently the same" criteria, (iv). Note that Funkhouser also formulates multiple realization in terms of his account of realization, rather than in terms of taxonomic differences as we do. So it may be that our accounts differ more than we here suggest, after all.

[8] In practice there are usually other differences between mechanical and automatic movements, e.g., to prevent over-winding. But we will ignore those in this discussion.

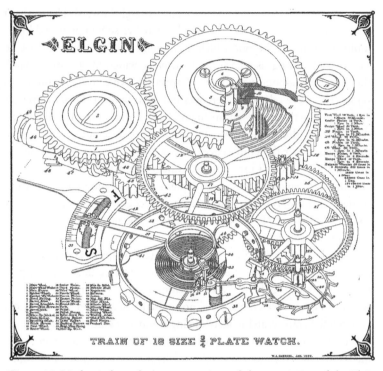

Figure 4.1 Mechanical watch movement. Artwork by permission of the Elgin Area Historical Society.

continuously until the energy is expended. Therefore some sort of mechanism for controlling the release of the energy is needed. This is the escapement, which stops the spinning of the gears and allows the energy to "escape" in periodic increments. This is normally achieved by the movement of a balance wheel, itself driven by a second spring, the balance spring or "hairspring." The period of the hairspring determines the rate of movement of the hands by periodically releasing the escapement, so that the rotation of the hands corresponds to the passing of time.

In contrast, quartz movement watches store their power chemically in a battery and release it electrically. The released electrical energy does two things: It powers a motor, which turns the gears and thus the hands of the watch; it also electrifies a quartz crystal that is tuned to oscillate at a specific frequency. The quartz oscillation is detected electrically and used to control the rate of movement of the motor, and thus of the hands.

Let us suppose that we have three watches. The first, *slow*, is a mechanical watch that runs slow. Second, *fast* is a mechanical watch that runs fast. Finally, *quartz* is a quartz movement watch that keeps the correct time. And let us suppose, only for the sake of this example, that we are behaviorists or "liberal" functionalists about watches: All three watches have the same function, viz., that of telling time. In that case we could say that *slow* and *fast* are not quite as good in performing this function as is *quartz*. And we can explain the difference in *slow* and *fast*'s performance on the basis of a slight difference in the properties of a component they share. The most common reason for otherwise similar mechanical watches to run fast or slow is that the period of the balance spring is incorrect because the spring itself is too long or too short. So let us stipulate that this is the only difference between *slow* and *fast*.

On the other hand, although the time-keeping properties of *quartz* are very similar to those of *slow* and *fast*—indeed, they may be more similar to each of *slow* and *fast* than are the time-keeping properties of *slow* and *fast* to each other—the mechanism in *quartz* is quite dissimilar from the mechanisms in *slow* and *fast*. Watchmakers consider *quartz* to be an altogether different kind of realizer of watch than *slow* or *fast*. The differences between quartz watches and mechanical watches are relevant to the performance of their watch function, and these differences also contribute to what they have in common with other watches. So quartz and mechanical watches seem to be multiple realizations of watch: They are the same in function but differ relevantly in realization, and these differences in their realization explain how they manage to perform the same function. They are thus "samely different" and "differently the same."

What about *fast* and *slow*—are they distinct realizers of watch? We think not. *Slow* and *fast*, despite differing slightly in their time-keeping capacities, keep "pretty" much the same time, and their physical similarity to each other is very close compared to their physical similarity to *quartz*. As the brief explanation above shows, because *fast* and *slow* are both mechanical watches they tell time in one way, whereas *quartz* tells time in a very different way.

A watchmaker would regard the small difference in time-keeping properties that *slow* and *fast* exhibit as merely an individual difference rather than a feature of taxonomic significance for watch movements. If the criteria are arbitrarily fine-grained, then no two instances of any kind are exactly similar.[9] Accordingly, a science of watches that distinguishes *slow* from *fast* strikes us as both methodologically unsound and anomalous.

But the reason to prefer the judgment that *slow* and *fast* count as just one kind of realizer is not merely that the difference between them is small. Rather, it is justified by consideration of the explanation for their functional differences. Why does *slow* run slow and *fast* run fast? We have assumed that these differences trace to a single difference in their components: The length of the hairspring in each. This fact gains importance when one considers the motivation for distinguishing *slow* from *fast*. The idea was that the two are different realizers of *watch* because they have different capacities—one is a fast timekeeper and the other a slow timekeeper. However, if, as we have assumed, these differences can be fully explained by their physical difference, then the verdict begins to look dubious. The entire *physical* difference between the two watches corresponds to the *functional* difference between them. Thus, to the extent that they are physically similar, they function similarly; and to the extent that they differ functionally, they differ physically. This suggests to us that *slow* and *fast* realize time keeping in the same way. The differences they exhibit in their functional capacities are simply the consequence of an individual difference in their mechanisms. What they "do" the same, they realize the same; and what they "do" differently, they realize differently. Their differences are relevant differences, but they are differences that explain difference, not differences that explain sameness.[10]

In contrast, the differences and similarities between *quartz*, on the one hand, and *slow* and *fast*, on the other, require a distinct sort of

---

[9] Maybe this is not strictly speaking true for subatomic particle kinds. But that will not have a bearing on our questions about minds and brains.

[10] Suppose that rather than slow and fast, we consider two mechanical watches that keep accurate time but differ in the length of their hairsprings. This could be the case if other watch components were adjusted appropriately, such as adjustments to the balance or escape. According to Ken Aizawa (2013) this would be an example of multiple realization by "compensatory adjustments." Aizawa gives the example of electrical resistance in a wire, "given by the equation $R = l\,\rho/A$, where l is the length of the wire, ρ is the resistivity of the material out of which the wire is made, and A is the cross sectional area of the wire" (2013: 73). Resistance is a dependent variable determined by the three right-side variables, l, ρ, and A. The same value for R can be achieved (at least in theory) by infinitely many combinations of l, ρ, and A. On our view, however, these are not multiple realizers of resistance, they are all resistors in the same way.

explanation. It is true that the individual differences in the time-keeping capacities between *quartz* and the pair of analog watches also result from a physical difference between the three watches. However, unlike *slow* and *fast*, whose functional similarity owes to their physical similarity, the functional similarity between these two watches and *quartz* does not. *Quartz* doesn't function like *slow* and *fast* in virtue of its physical similarity to them: It functions like them in virtue of very different physical properties.

We frequently illustrate our account of multiple realization using artifacts, such as corkscrews and watches. We turn to these examples because multiple realization is much less common in naturally occurring systems than is usually recognized. The conditions for the kinds of variation that count as multiple realization are easy to create at a workbench or drafting table, but they are relatively rare in nature. Later we will argue that this is not surprising. For now, however, we want to point out that our account does not render naturally occurring multiple realization impossible. Indeed, we think there are some relatively clear examples. As we said earlier, we believe that there is multiple realization of eyes.

As Joram Piatigorsky notes, the distinguishing characteristics of eyes are themselves a matter of dispute. An eye might be "a photoreceptor that responds to light in some fashion or . . . a complex organ used in vision" (Piatigorsky 2008: 403–4). Many eye researchers distinguish eyes from simpler light detectors by their ability to form and transmit an image, which requires that the light-sensitive cells have some sort of "optical geometry" (Nilsson and Pelger 1994). Plainly it is the multi-part image-forming organ that fascinates vision scientists. It is this sort of thing of which Darwin famously said, "to suppose that the eye, with all its inimitable contrivances for adjusting the focus for different distances, for admitting different amounts of light, and for the correction of spherical and chromatic aberration, could have been formed by natural selection, seems, I freely confess, absurd in the highest possible degree" (1859: 186). But Darwin, as we noted earlier, immediately goes on to argue that eyes have indeed evolved. And if Michael Land and Russell Fernald are right, different image-forming eyes have evolved multiple times (Figure 4.2):

At last count, there were ten optically distinct ways of producing images [Figure 4.2, Left and Right]. These include nearly all those known from optical technology (the Fresnel lens and the zoom lens are two of the few exceptions that come to mind), plus several solutions involving array optics that have not really been invented. Some of these solutions, such as the spherical graded-index lens [Figure 4.2, Left e], have evolved many times; others, such as the reflecting superposition eyes of shrimps and lobsters [Figure 4.2, Right j], have probably only evolved once. (1992: 7)[11]

Among the ten basic kinds of eyes, the biggest distinction is between "simple" or "camera" eyes (Figure 4.2, Left) and compound eyes (Figure 4.2, Right). We think that this distinction, even leaving aside the further variations within the classes of simple and compound eyes, provides a good example of multiple realization according to our recipe.

Let us ask the specific and contrastive question: Is the kind *eye* multiply realized by simple and compound eyes? The first thing we need to determine is whether there is a taxonomic class to which all of these things belong. Yes, they are all eyes; in fact, as discussed above, they are all specialized organs that form images. Criterion (i) is met. Second, we must show that there is some other taxonomic schema that distinguishes simple from camera eyes. And as we have already described, there is. Criterion (ii) is also met. Now we need to know whether the differences observed between simple and camera eyes are relevant differences—are they differences that are relevant to their being eyes. Again, the answer is yes. Image formation is central to being an eye, and "the mechanisms involved [viz., simple and compound eyes] really are very different and represent topologically 'concave' and 'convex' solutions to the problem of image formation" (Land and Fernald 1992: 8). The differences in the optical geometry of simple and compound eyes are relevant differences, so criterion (iii) is met. Finally, we ask, do those differences contribute to the sameness of simple and compound eyes, or only to their differences? And the answer, already indicated, is that simple and compound eyes are different ways of doing the eye's image-forming task: Using concave and convex arrays of receptors, with single and multiple focusing systems, with images integrated on a receptor surface or at later stages of processing.

---

[11] Later Land and Fernald write: "Eyes with well-developed optical systems evolved many times at the end of the Cambrian period. There are now about ten optically distinct mechanisms. These include pinholes, lenses of both multielement and inhomogeneous construction, aspheric surfaces, concave mirrors, apposition compound eyes that employ a variety of lens types, and three kinds of superposition eye that utilize lenses, mirrors, or both" (1992: 25).

Simple and compound eyes are not just differently different, they are differently the same. Criterion (iv) is met.

Of course, whether two things belong to one kind or not, within a given science, is often a difficult question for investigators, and may depend as much on the explanatory and methodological utility of the kind or the science as on any criterion for kind membership that might

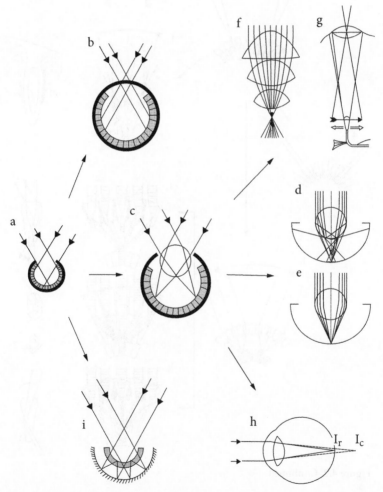

Figure 4.2 Kinds of eyes (Land and Fernald 1992). Modified with permission from the *Annual Review of Neuroscience*, Volume 15, ©1992 by Annual Reviews, <http://www.annualreviews.org>.

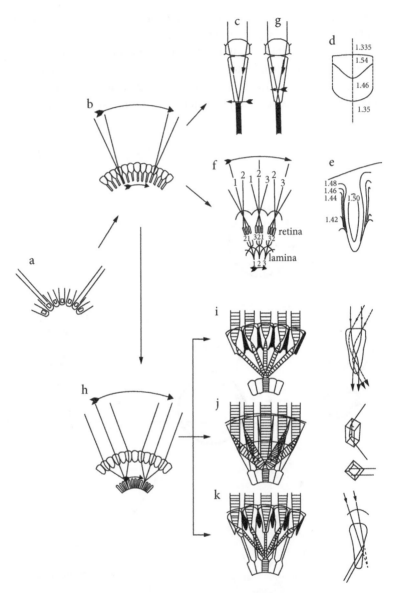

Figure 4.2 Continued

be fixed in advance. Sameness and difference are relative—investigators must decide, given their particular and often various purposes, when judgments of sameness and difference are merited (Shapiro 2000; Polger 2008, 2009a, 2009b; Shapiro and Polger 2012). This is not to deny that sameness and difference are objective matters. On the contrary, the interesting fact is that there are many objective similarities and differences in the world, and that different sciences concern themselves with different objective similarities and differences. That is why questions about multiple realization must always be specific and contrastive. If the question about multiple realization were just, "Are there differences among corkscrews?" or "Are there differences among eyes?" then the answer is yes: Corkscrews are similar to one another in indefinitely many ways, and dissimilar from one another in indefinitely many ways. So too for watches, eyes, pains, and brains. After all, everything is similar to everything else in indefinitely many ways, and dissimilar from everything else in indefinitely many ways.

3 What Multiple Realization Is Not

We have argued that multiple realization is not just variation, it is a distinctive kind of variation. This is why the mere fact that the world is Heraclitean—that all is flux—does not by itself show that mental states are multiply realized. Multiple realization requires a special pattern of variation: Relevantly the same function performed in relevantly different ways, where the differences contribute to the sameness in function and not just to the differences in function.

Returning now to the topic of minds, we submit that the phenomenon of interest is not the multiple realization of mental processes *simpliciter*, or the possibility that they might in some way or other be multiply realized. We're interested in the specific question of whether mental processes are multiply realized by brain processes, or by physical kinds that include both brain processes and non-brain processes. According to our approach, this amounts to the question of whether the taxonomy of mental processes provided by scientific psychology corresponds to the taxonomy of brain processes provided by the brain sciences. Do they carve the world at the same joints, or not?

When philosophers talk about multiple realization without qualification, we can only assume that they are talking about the multiple realization of mental processes by brain processes that meets the conditions we have described. That is the phenomenon, if it occurs, that could at least potentially present an obstacle to brain and identity theories. And if so, that would be a mark in favor of functionalist or realization theories, as Putnam, Fodor, and many others have concluded.

But just as not any variation in composition makes for multiple realization of corkscrews, not any variation in brains makes for multiple realization of mental processes by brain processes. Consider Ned Block's story about tiny people who fly about in little spaceships mimicking the behavior of atoms (1978). Suppose, probably contrary to fact, that physicists decided that those things would be, in fact, atoms—that is, suppose that, against all odds, the best overall physical theory counted the tiny piloted spaceships as members of the kind *atom*, as fulfilling what we could call the atom function. If so, we assume that would count as a case in which atoms are multiply realized: The atom job can be done by more than one kind of thing.[12] Even granting the stipulation that atoms are multiply realized in this story, there is no reason whatsoever to think that mental states are thereby multiply realized, or even that brain states are thereby multiply realized. As far as this story goes, if the little flying spaceships really are atoms according to physics, then whatever those atoms do they all do in the same way—to wit, in the atom way. They make up molecules, which make up proteins, and so on until we get to cells like neurons that make up brains. And those brains do their brainy jobs in just the same way whether their atoms are made of fundamental particles or tiny spaceships.

Block's example is definitely fanciful and probably impossible. We rather doubt that atoms are functional kinds, to begin with. We consider the case merely to illustrate as starkly as possible a phenomenon that we think is common but often overlooked. Simply put, sciences are not required to make all and only the same distinctions as one another. If the science of the components of atoms says that there is a difference between things made of fundamental particles and things made of tiny flying spaceships, it does not automatically follow that the sciences of

---

[12] If the science of subatomic particles is insensitive to the difference between fundamental particles like electrons and obviously non-fundamental "particles" made of tiny people flying tiny spaceships, then we shall have to come up with a different example. So assume, for the sake of argument, that they do count as different "ways" of doing the atom job.

things that depend on those things must also make that distinction. Unlike Darwinian variation, compositional variation is not inherited.

This is of course not to say that variation *never* bubbles up from parts to wholes. Of course it does. In the common vernacular of supervenience, when what is realized supervenes on what realizes it, the only way that there can be variation in what is realized is when there is variation in the realizers. We suspect that there are even some very interesting cases of this kind of ramification of variation, such as "level skipping" where variations in a realizer make no difference to what it realizes, but still makes a difference to what is realized by what it realizes. But this does not always happen—such ramification is not automatic. We encountered this earlier when noting that differences in in human and octopus retinas do not automatically make them different realizations of the kind *eye*— or even of the kind *camera eye*.

The idea that the world is Heraclitean is related to the idea that the world, as Bill Lycan describes it, is "functions all the way down" (1987: 47). We take Lycan to believe that the world also exhibits multiple realization "all the way down." But that's not true—it is not the case that there are no type identifications between any explanatory taxonomies, all the way down. Examples are abundant (if controversial): Lightning and electrical discharge, water and H_2O, gold and the element with atomic number 79, temperature (in a gas) and mean molecular kinetic energy, and more. This demonstrates that even if functional kinds persist near the most fundamental levels of nature, we shouldn't expect multiple realizability to be pervasive at every level. If, as is imagined in Ned Block's story, it turns out that *atom* is multiply realizable, it doesn't follow that *brain*, *sensation*, or *belief* are as well. The lesson is that there is no shortcut from some general metaphysical principles to the truth of the multiple realizability of mental processes by brain processes (Polger 2004a, 2011). And, really, we should expect that because multiple realization is an empirical claim. What we need is evidence.

The evidence we're looking for is not just that of a few cases, here and there. That would not be enough to give the upper hand to functionalist and realization theories, as multiple realizability is supposed to do. Examples of multiply realizable psychological processes must be substantial and important in the overall theoretical model of the mind sciences. In particular, they have to give us reason to think that identities between psychological kinds and neuroscientific kinds are sufficiently rare or unimportant that they do not play a substantial and important role in our total understanding of minds and brains. Anything less than this sort of evidence leaves the advocate of multiple realization with only a weak or speculative case against the brain state theory.

Naturally, the best evidence for the truth of multiply realizability would be evidence of actual multiple realization. This would be evidence of psychological processes and brain processes that in fact have a pattern of variation that satisfies criteria (i) through (iv) of the Official Recipe. The next best evidence would be empirical support for a total model of minds and brains that implies that psychological processes are the kinds of things that can have a pattern of variation that satisfies criteria (i) through (iv), even in the absence of actual cases.

To the evidence we now turn.

PART II

The Evidence for Multiple Realizability

5

Evidence for Multiple Realization
Neural Plasticity

1 Evidence and Ambition

As we've noted in previous chapters, Hilary Putnam set out a challenge for identity theorists. Insofar as both functionalism and the identity theory were intended as empirical hypotheses, Putnam took evidence to be relevant to their assessment. And on the basis of evidence he believed functionalism held the upper hand. "[I]t is at least possible," Putnam said, "that parallel evolution, all over the universe, might *always* lead to *one and the same* physical 'correlate' of pain. But this is certainly an ambitious hypothesis" (1967 in 1975: 436).

Because Putnam emphasized the empirical nature of his functionalist hypothesis, one might expect that he would have said something about the evidence he takes in its support—the evidence for multiple realization. What did he see in mammalian and mollusk behavior and anatomy that led him to conclude that pain, in these organisms, is multiply realized? Does it just stand to reason that pain in organisms as various as otters and octopuses would be realized differently? One might have thought the same about mammalian and mollusk eyes but these, Putnam concedes, are realized similarly. As should by now be obvious, judgments regarding the similarity of psychological states and the difference in would-be realizers rest on many considerations that Putnam did not anticipate. When we distinguish between multiple realization and other sorts of variation in nature, perhaps it is multiple realization that appears to be an "ambitious" hypothesis.

2 Early Approaches to Evidence for Multiple Realization

We argued that multiple realization is not just any sort of variation in nature; it is a particular pattern of variation. And Putnam seems to agree, for he acknowledges that human and octopus eyes can be the same kind of thing despite their many differences—human eyes have "inside-out" retinas and octopus eyes do not. Although Putnam seems to think we have evidence of the special kind of variation between minds and brains that satisfies the requirements for multiple realization, he says little about what he takes that evidence to be. It appears that he thought that the multiple realization thesis is so obvious he felt no need to delve very deeply into a discussion of evidence. And this complaisance seems to have gone unquestioned in the years following publication of Putnam's hypothesis. But to take for granted that obvious variation is evidence for multiple realization is to neglect the subtlety of the phenomenon. Putnam should have been more circumspect.

To our knowledge, Ned Block and Jerry Fodor (1972) were the first to discuss at any length the sorts of evidence that might favor the multiple realization hypothesis over the identity theory. Block and Fodor begin their case by observing that the "argument against physicalism [what we've been calling the identity theory] rests on the empirical likelihood that creatures of different composition and structure, which are in no interesting sense in identical physical states, can nevertheless be in identical psychological states" (1972: 160). And they are confident that psychologically similar—identical, even!—states might be present in creatures that are not physically identical in any "interesting sense." But as we have explained, much of the debate about multiple realization must turn on questions about whether there is evidence of creatures that share psychological states and processes but are not physically similar in any "interesting" ways. And we assume that whether two things have anything "interesting" in common is not merely a matter of taste. For example, we assume that it is at least sufficient for a similarity to count as "interesting" if it plays a role in some actual explanation or experimental procedure. So we understand Block and Fodor's claim about "interesting" similarities to be a claim about scientific practices and explanations. Is it entirely impossible for *some* scientific domain to recognize similarities in the would-be realizers of the psychological states and processes that Block and Fodor have in mind? We are skeptical.

Consider again the case of the camera eye. We do not deny that the cells composing octopus and mammalian camera eyes differ. But biochemical differences between the eyes of octopuses and mammals do not prevent them from belonging to the same category from the perspective of some other science. The kind *eye*, from the perspective of visual optics, is realized in the same way in octopuses and mammals despite structural and cellular differences. When making taxonomic judgments, claims about *the* interesting sense of physical sameness must be made with caution. Practitioners of one scientific discipline may not attend to the "interesting" ways in which two samples are similar, while those of another scientific discipline do. Hence, we can agree that relative to one science there are no interesting similarities between would-be realizers of psychological kinds, while still maintaining that relative to some other science the would-be realizers are in fact quite similar. The devil is in the details, so one needs to examine particular cases.

In the ensuing decades since Putnam's empirically speculative assertion of multiple realization, philosophers have pursued two distinct lines of evidence. On the one hand, some advocates have purported to offer actual examples of multiple realization. That is, they have provided case studies that are thought to exhibit the pattern of variability in the world that is distinctive of multiple realization. This seems to be what Putnam took himself to be offering—even if, as we have argued, he failed to mount much of a case. Call this the *direct evidence* for multiple realization. On the other hand, some advocates have opted to skip over case studies and appeal instead to general considerations. Rather than looking to examples to infer the prevalence of multiple realization, they argue that overall evidence supports hypotheses that are compatible with multiple realization. On that basis, the line of thought goes, we should accept the possibility of multiple realization, never mind a lack of actual examples. Call this the *indirect evidence* for multiple realization. Block and Fodor take this latter route. In this chapter we examine direct evidence for multiple realization. In the next, we tackle Block and Fodor's indirect evidence.

3 *Situs Inversus Viscerum*

As we mentioned in earlier chapters, philosophers have thought that cases of neural plasticity constitute actual examples of multiple realization. Plasticity is the capacity of brains and neurons to reorganize themselves, often to compensate for damage to some brain areas or neural connections. For instance, memory, attention, or perception functions that once were realized in cortical areas X, Y, or Z, might after trauma or training be realized in adjacent or connected areas X*, Y*, or Z*. A blind person's visual cortex might process auditory stimulation (Klinge et al. 2010; Wong and Bhattacharjee 2011). And a blind person might train her visual cortex to process auditory information, such as those resulting from verbal clicks, endowing her with a rudimentary sight (Thaler et al. 2011). Work with sensory substitution devices has revealed that tactile information, coming from the tongue or the back, can cause activation in visual cortex, leading to visual experience in blind subjects (Bach-y-Rita and Kercel 2003). These findings, undoubtedly remarkable, may nevertheless not make the watertight case for multiple realization that they *prima facie* suggest. To see why, let's first examine a case that, although not involving neural plasticity, is similar in spirit. We offer the following example as a way of framing the issues concerning the evaluation of direct evidence for multiple realization.

Roughly one in 10,000 human beings have a condition known as *situs inversus viscerum—situs inversus* for short. In these people, organs normally on the left side of the body appear on the right, and those on the right appear on the left. Thus, a person with *situs inversus* will have her heart, spleen, and stomach on her right side, and her appendix, gall bladder, and liver on her left. Of course, this rearrangement entails as well that all nerves and blood vessels associated with these organs also be inverted. The anatomy of a person with *situs inversus* thus differs dramatically from that of a statistically normal person. Should an individual with *situs inversus* require a heart transplant, for instance, special difficulties ensue, for a normal left-side heart cannot simply be plugged into a body that's "wired" for a right-side heart.

Suppose now that one were to claim that the realizers of organs in a *situs inversus* person differed from those in a normal person. The argument would go like this (*mutatis mutandis* for each affected organ):

1. The realizer of *heart* in a normal person is on the left side.
2. The realizer of *heart* in a *situs inversus* person is on the right side.
3. Left-side hearts differ from right-side hearts.
4. Therefore, the kind *heart* is multiply realized in normal and *situs inversus* persons.

One might buttress the argument with the observation that left-side and right-side hearts cannot simply be swapped for each other, as the discussion above about transplants noted. The hearts differ not simply in their location, but in their construction.

On first appearances, the organs of a *situs inversus* individual satisfy the recipe for multiple realization that we have adopted. First, they are the same in function: For each organ that does F in a normal person, there is an organ that also does F in a *situs inversus* person. Indeed, it is for this reason that *situs inversus* so often goes unnoticed. Unless a person with *situs inversus* seeks treatment for an unrelated injury or illness, she may never know that her organs mirror those of a normal person. But, on the other hand, as the supporter of the argument above may insist, the *situs inversus* organs do differ. That is why organ transplants between normal and *situs inversus* individuals present unusual challenges.

Situs inversus organs (henceforth *SI-organs*) and normal organs (*N-organs*) are similar in function. A physiologist, for instance, might not distinguish them. But they differ in ways that a transplant surgeon would note. Same but different. Multiple realization! But we think this line of reasoning rests on a mistake.

The reason to deny that, e.g., an SI-heart is a different kind of realization than an N-heart rests on the same considerations that we introduced in the previous chapter with our discussion of corkscrews and watches. Imagine for a moment that a science of corkscrews were to develop. Corkscrew scientists would face the problem of taxonomizing the variety of corkscrews in the world. Like any taxonomic task, the corkscrew scientists would need to decide on which properties of corkscrews to focus—which properties of corkscrews mark a kind-difference and which do not? This question is just like those scientists in other fields must ask: Which features of atoms distinguish hydrogen from oxygen? What properties distinguish mammals from fish? When is a rock sedimentary and when igneous?

It is tempting to think that SI-organs and N-organs are different realizers. But we suggest that the difference between SI-organs and N-organs is irrelevant. The differences between SI-organs and N-organs are not differences that figure in identifying what they are—hearts, livers, lungs, and so forth. And they are not differences that require that we employ different explanations either for why these organs evolved or what they currently do in those creatures that have them. Of course there could be a different developmental or genetic explanation for how SI-organs come to be. But that explanation will parallel a similar explanation for how N-organs develop. The difference between SI-organs and N-organs is dramatic. But even this is not a difference that makes a difference in kinds of organs, just as the difference between red and blue handles is not a difference that makes a difference in kinds of corkscrews. Or, more to the present point, right-handed and left-handed tools—or for that matter, between right-handed and left-handed people.[1]

We've chosen to introduce the issues surrounding multiple realization and neural plasticity with discussion of *situs inversus* because, we submit, the argument for multiple realization based on observations of neural plasticity is susceptible to the same kind of criticism that we just made against the argument for multiple realization based on *situs inversus*. Not all differences in would-be realizers are differences that ought to amount to multiple realization. Some differences matter, but some do not. We claim that the differences between SI-organs and N-organs do not.[2] The answer to the specific question, "Relative to the taxonomic practices of physiology, is *heart* (and, mutatis mutandis for the other organs) realized one way by an SI-heart and another by an N-heart?," is no. Likewise, once the questions about neural plasticity are made more specific, we believe that many examples of neural plasticity do not justify the conclusion that psychological states are multiply realized.

---

[1] We would say the same thing about another familiar form of polymorphism, viz., sexual dimorphism, the phenotypic difference between males and females in many species.

[2] For instance, the textbook diagrams that trace the paths of activity in each organ will describe the same order of operations for SI-organs and N-organs.

We are now in a better position to evaluate the evidential weight of neural plasticity for multiple realization. Neural plasticity is evidence for multiple realization only if it reveals the presence of multiple neural structures that perform the same function and that differ in taxonomically relevant ways. But, as the discussion of *situs inversus* suggested, we cannot take for granted that even large or obvious differences will turn out to be taxonomically relevant differences. In the previous chapter we defended our Official Recipe for multiple realization. There are at least four jointly necessary conditions for multiple realization to occur:

(i)   As and Bs are of the same kind in model or taxonomic system S1.
(ii) As and Bs are of different kinds in model or taxonomic system S2.
(iii) The factors that lead the As and Bs to be differently classified by S2 must be among those that lead them to be commonly classified by S1.
(iv) The relevant S2-variation between As and Bs must be distinct from the S1 intra-kind variation between As and Bs.

According to us, some big and obvious differences among things fail to satisfy the Official Recipe.

Consider how these four conditions apply to the case of *Situs Inversus Viscerum*. Either the difference between SI-hearts and N-hearts is a difference in kind or it is not. If not, then SI-hearts and N-hearts are not different realizers of heart. Criterion (ii) thus fails to apply. But suppose, for the sake of argument, that the difference between SI-hearts and N-hearts does mark a difference in kind, after all. Maybe due to the direction of the Earth's magnetic field N-hearts pump better than SI-hearts and for this reason physiologists might view N- and SI-hearts as distinct kinds of organs. Still, the features of an SI-heart that endow it with the capacity to pump blood at all are the same features that explain an N-heart's capacity to pump blood. And, whatever differences exist in the capacities of the SI- and N-organs occur in virtue of corresponding differences in their features. Quite simply, even if SI- and N-organs are, relative to the taxonomic practices of physiology, different organs, they are not differently the same. That is, they do not perform the same function in different ways. Rather, they are *samely* the same, and differently different. That is, what they do the same, they do in the same way, viz., they pump blood in virtue of similar processes of contraction and expansion. And, what they do differently, they do differently because of their differences—in virtue, say, of the direction in which they force blood flow. Hence, although (ii) is satisfied under this second supposition, (iii) and (iv) are not. We believe that this is typical of prima facie cases of multiple realization, including many involving neural plasticity. Let us see.

4 Neural Plasticity and Multiple Realization

Many philosophers take the various phenomena of neural plasticity to provide direct evidence of multiple realization. There are many forms of neural plasticity. One important sort is *synaptic plasticity*, the adjustment of synaptic connections in response to use or disuse (Buonomano and Merzenich 1998). But synaptic plasticity by itself is not what philosophers who advocate for multiple realization usually have in mind, for it is not an example of same-but-different. Instead, synaptic plasticity is an example of different-and-different—changes in the synaptic structure (e.g., expression of new neurotransmitter receptor sites) correspond to changes in the neural processing (Polger 2009a). Instead, fans of multiple realization usually think of plasticity in terms of phenomena that we call cortical functional plasticity, wherein whole areas of cortex seem to perform different tasks at different times or in different subjects. Let us consider two examples of cortical plasticity, which we can think of as representative of common and radical varieties of neural plasticity.

A common sort of neural plasticity affects the size, shape, and arrangement of the areas in somatosensory and motor cortex that receive projections from various parts of the body. Area 3B of the primary somatosensory cortex in an adult monkey contains a map-like representation of the monkey's hand (see Figure 5.1a), with portions of the map corresponding to the pads and digits of the hand (Figure 5.1b and 5.1c). The portion of the map that represents stimulation to the ventral surfaces of the thumb and first two digits receives signals via the median nerve. When this nerve is severed, the organization of the somatotopic representation of the hand changes. Areas of the map originally devoted to representing stimulation to the thumb and first two digits no longer receive input (Figure 5.1d). After time, however, the deprived regions of area 3B begin to receive information from the radial nerve, which connects to the dorsal rather than the ventral surface of the monkey's hand (Figure 5.1e). Thus, primary somatosensory cortex displays plasticity. Damage to the median nerve

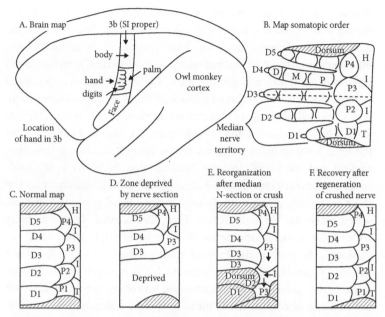

Figure 5.1 (A) Area 3B of the somatosensory cortex devoted to representation of the hand. (B) The ventral portion of the hand represented in area 3B. (C) The somatosensory representation of the ventral portion of the hand. (D) The effect that severing the median nerve has on the representation of the hand. (E) The somatosensory cortex after about one month of recovery. Areas D1, D2, and D3 are now innervated by inputs from the dorsal surface of the hand via the radial nerve. D1–D3 now represent dorsal portions of the first three digits and pads of the monkey's hand. (F) Normal cortical map recovered (Kaas 1991: 143). Modified with permission from the *Annual Review of Neuroscience*, Volume 14, ©1991 by Annual Reviews, <http://www.annualreviews.org>.

results in reorganization of the cortex. Regions of area 3B that once represented ventral surfaces of the thumb and first two digits of the monkey's hand come to represent, via the radial nerve, dorsal surfaces of these digits.

The plasticity this case illustrates is quite typical of sensory and motor cortexes.[3] Often following peripheral damage there follows a period of

---

[3] Richardson (2009) discusses this sort of plasticity in language-processing areas. He concludes that they provide examples of "same but different"—our criteria (i) and (ii)—and he takes that to be sufficient for multiple realization. So he is not evaluating plasticity

dormancy in the areas of the cortex that had once been innervated by the damaged areas. Then, areas nearby the damaged ones "expand" into the quiescent area of cortex, just as the dorsal portions of the monkey's hand expanded into areas of 3B that had once received activation from the ventral surfaces of the thumb and first two digits. Cortex that had once processed ventral stimulation now takes on a new task: Processing dorsal stimulation.

A more radical variety of plasticity is illustrated by the "rewired" ferrets studied by Mriganka Sur and his colleagues. The standard pathway from the ferret's eye to its visual cortex travels through the lateral geniculate nucleus and the lateral posterior nucleus. Von Melchner et al. (2000) redirected retinal axons from the right visual field that usually project to these areas, connecting them instead to the medial geniculate nucleus, which innervates the audio cortex. Thus, the "rewired" ferret's auditory cortex received visual information from the ferret's right visual field.[4]

Making this study particularly interesting is the fact that the visual and auditory cortexes display very different kinds of organization. Within visual cortex, groups of cells are arranged into orientation columns, with each column of cells especially tuned to specific orientations of a stimulus (Sharma et al. 2000). Moreover, the visual cortex contains a 2D map of the retina, with each point on the retina corresponding to a point on this map. In contrast, the auditory cortex contains no columns of orientation-sensitive cells. Nor does it contain a 2D map of auditory space. Rather, the cochlea maps onto a 1D map in auditory cortex and neurons in auditory cortex are grouped in clusters that receive excitation from both ears, or excitation from one ear and inhibition from the other (Roe et al. 1990). Given the magnitude of these differences in visual and auditory cortexes, it would be quite surprising if the rewired ferrets regained vision in their right visual fields.

But this they did! The ferrets that had been trained to respond in one way to an auditory stimulus and another to a visual stimulus displayed the visual response to stimuli presented to their right visual fields (von Melchner et al. 2000). Rewired ferrets were also tested for their visual

against our criteria (iii) and (iv). See also Barrett (2013) for related considerations, including arguments that some cases of plasticity do satisfy our criteria (iii) and (iv).

[4] Each eye has a right and left visual field. Processing of information from the right visual field of each eye takes place in the left hemisphere; information from the left visual field is processed in the right hemisphere.

acuity, and they were able to discern gratings of various frequencies and at various contrasts. In short, the rewired ferrets appear able to see with their auditory cortex.

Given the prominence of neural plasticity in discussions of multiple realization, one might suppose that evidence of the sort described above makes a clear and convincing case for the multiple realization of visual perception. If we allow that the rewired ferrets process visual information in the same way that normal ferrets do, and that the auditory cortex that realizes visual processing in the rewired ferrets does indeed differ from visual cortex, then, or so it seems, we have a legitimate case of "same but different." Same visual capacities; different cortical realizations. But not so fast. Closer inspection of the examples tells a different story.

Take first the common example of neural plasticity, such as the recruitment of somatosensory cortex for processing tactile information from the monkey's hand. Far from showing multiple realization, which involves sameness at the higher level and difference at the lower, this example instead illustrates something like the opposite—difference at the higher level despite sameness at the lower level. When the median nerve that carries information from ventral portions of the monkey's hand is severed, the subsequent reorganization of somatosensory cortex does not end up realizing in a different way the same psychological function. Instead, regions of somatosensory cortex that once processed information from the ventral regions of the hand subsequently process information from dorsal regions of the hand. Thus, rather than the same function being realized by different neural organizations, we end up with different functions (dorsal rather than ventral mapping) being realized by the same neural organization. That is, the same regions of somatosensory cortex that once realized part of a map of the ventral portion of the monkey's hand comes to realize part of a map of the dorsal portion.

This kind of neural plasticity, in which a single kind of neural substrate goes from realizing one sort of psychological function to realizing a different sort, provides no evidence for multiple realization. Rather than exhibiting the same cognitive process being performed in different neural ways, it is an example of one brain structure doing different tasks. Michael Anderson has emphasized the extent of neural reuse and multifunctionality, and even the simultaneous realization of multiple processes by shared neural realizers (2010, 2014). Multifunctionality could be a problem for identity theorists, but multifunctionality is not evidence of multiple realization; it is a different pattern of variation than the one we articulated in our Official Recipe.[5]

What about the more radical and unusual case, in which auditory cortex comes to realize visual processing? This appears more promising as evidence for multiple realization.

Consider first, before examining the case in any more detail than already presented, just how tremendously unlikely it is that the auditory cortex in the rewired ferrets can actually realize normal visual processing. We've noted that visual cortex displays a specialized architecture involving columns of orientation-sensitive cells. Moreover, its cellular layout provides resources for a 2D map of the visual world. In contrast, auditory cortex lacks both these features. One must wonder, then, how visual processes could be realized in a structure that, speaking loosely, was designed for processing of a very different sort. We have no expectation that the processes that a digital calculator employs to calculate multiplication products could be realized in a clock or a microwave oven or a television; why should we think that the processing that occurs in visual cortex could also take place in auditory cortex? Surely the default assumption should be that it cannot; that a cortex designed for auditory perception cannot also realize visual perception.

But, of course, this is the assumption that advocates of multiple realization believe to be false—and not just in exceptional circumstances, but all the time. We should expect, they think, that pretty much any psychological processing could be realized in pretty much any kind of substance. Instances of neural plasticity like that involving the rewired ferrets turn out not to be surprising after all given the commitments of functionalism; only the identity theorist would express surprise at such results.

However, any hope that the plasticity on display in the rewired ferrets supports multiple realization depends on a very superficial examination of the case. The criteria we have offered for identifying legitimate

---

[5] Multifunctionality highlights that the operating units of brains are probably not localizable chunks, and thus that the mind-brain identity theorist should not be interpreted as identifying mental states with portions of brain that could be scooped out with a melon baller. We usually talk about the identification of mental process kinds with brain process kinds, and we appeal to neuroscientists to explain what sort of thing—or sorts of things—brain processes may be. Sensations are brain processes, on our view; but a lonely c-fiber firing is not a very realistic candidate for the process of experiencing pain. C-fiber firing might be part of any number of brain process, but it is not the whole of any sensory or cognitive process.

examples of multiple realization reveal why. With respect to the rewired ferret, the first two criteria require that the visual capacities of the rewired and normal ferrets be the same while the auditory cortex in the rewired ferret differs from the visual cortex in the normal ferret. Satisfaction of these criteria accounts for the "same but different" characteristic of multiple realization. Let's consider first the "sameness" of the visual capacities in rewired and normal ferrets.

If visual processing in the visual and auditory cortexes were indeed the same, we should expect the normal and rewired ferrets to perform identically in discrimination tasks. Yet rewired ferrets perform markedly worse in a number of tasks. For instance, von Melchner et al. exposed both groups of ferrets to grated patterns of bands, where both the contrast between the bands and the spatial frequency of the bands varied between trials (2000).

In all cases, the ferrets' discriminatory capacities increase as the contrast of the grating pattern increases and as the spatial frequency of the grating decreases. The experiment also showed that in the rewired ferrets' intact left visual field, performance is close to the normal ferrets'. But in the rewired right visual field, rewired ferrets show significant degradation in their discriminatory ability, being unable to detect gratings at lower contrasts or higher spatial frequency than the normal ferret.

Clearly, despite the uncontested fact that visual processing in the rewired ferrets is realized in auditory rather than visual cortex, this visual processing differs from that which occurs in the visual cortex, to the detriment of proper visual functioning.[6] Were the processing the same, we should find rewired and normal ferrets performing roughly the same in the discrimination task. Consequently, looking to cases of neural plasticity like that in the ferret does not provide the support for multiple realization that a first glance might have suggested. Suppose, just to make the point more tangible, that one wrote an algorithm for multiplication and implemented the algorithm in calculator A. The algorithm works

---

[6] Alternatively, as suggested by an anonymous reviewer, we might take the functional differences between the visual cortex and the rewired auditory cortex to be akin to the differences in the two mechanical watches that we discussed in Chapter 4. We have doubts about this strategy; but note that its adoption would please the identity theorist, who would take the visual differences to be instances of individual variation in a single type of visual capacity, underwritten by individual variation between type-identical neural mechanisms.

just as it should, delivering the correct product to any pair of numbers. A friend now tries to compile the algorithm in a physically and structurally distinct kind of calculator, B. Calculator B will also provide correct answers to multiplication questions, except for those involving numerals with more than three digits. Surely this finding challenges the claim that the two calculators implement the same algorithm. The algorithm in calculator B must have undergone some change when "forced" to work with unsuitable hardware, and this explains why it produces outputs unlike those that the algorithm in calculator A produces.

So much for the "sameness" in the same but different analysis of visual capacities in normal and rewired ferrets. What of the difference? Not only does the case for multiple realization require that the ferrets possess the same kind of visual processing, but also that this processing is realized in different ways. Here the proponent of multiple realization might feel on safer ground. No one disputes that auditory cortex and visual cortex differ, and so *if* the visual capacities of normal and rewired ferrets were the same, the inference to multiple realization appears solid. However, the point with which we introduced this discussion now becomes relevant. How could something as distinct from visual cortex as auditory cortex possibly realize visual processing? Auditory cortex lacks the columnar organization of visual cortex and is not structured for purposes of representing a 2D space. That something so unlike visual cortex could behave just as visual cortex would be something like a small miracle.

The answer, it appears, is that auditory cortex reorganizes itself as a consequence of the retinal inputs it receives. What one notices when examining the auditory cortex of a rewired ferret is a cortex structured something like a normal auditory cortex, but also something like a normal visual cortex (Sharma et al. 2000). Thus, the rewired auditory cortex displays columns of orientation-sensitive cells, just as normal visual cortex does. It also contains regions that bear horizontal connections to each other, as do regions in visual cortex. But whereas one finds a great number of orientation maps in visual cortex that resemble something like a pinwheel, in rewired auditory cortex the density of these maps is far lower. Moreover, the volume of the regions of orientation-sensitive cells in rewired auditory cortex is much larger than the volume of these regions in visual cortex.

Described colloquially, the auditory cortex in the rewired ferrets appears to be trying its best to turn itself into a visual cortex for purposes of processing the information it receives from the retina. However, just as an attempt to make a clock from bicycle parts will get you only so far, so too the auditory cortex faces severe restrictions on how similar to a visual cortex it can become. As Sharma et al. observe, "[t]he differences between orientation maps and horizontal connections in rewired A1 and V1 suggest constraints on activity-dependent plasticity" (2000: 846). The intrinsic structure of auditory cortex limits how well it can accommodate visual inputs. Try as it might, it will never perform as well as normal visual cortex. Even if we think of these deficits as merely individual differences, our lesson holds: The normal and rewired ferrets have visual capacities that are similar insofar as their brains are similar, and different insofar as their brains are different.

We have seen that plasticity like that in the rewired ferrets fails the first condition for multiple realization. Visual cortex and rewired auditory cortex do not do the same thing. They do, on the other hand, satisfy the second condition, namely visual cortex and auditory cortex do differ. However, the observations above regarding the limitations on rewired auditory cortex suggest that the third condition for multiple realization, like the first, also fails. According to this condition, the differences between visual and auditory cortex—the differences that promise to fulfill the "different" part in the "same but different" analysis of multiple realization—must contribute to the sameness in their psychological function. That is, the two cortexes count as multiple realizations of visual processing only if their differences are different ways of realizing visual discrimination.

The motivation for this requirement, recall, is that if the cortexes implement visual perception similarly, then there is no obstacle to giving a common explanation for both in terms of the similar features of their realizers. The two waiter's corkscrews that differ in color must, *a fortiori*, differ in some physical respects. However, according to the taxonomic practices of imaginary corkscrew science, the physical differences between the two corkscrews do not necessitate distinguishing them in kind. Color differences are differences that make no difference, from the perspective of corkscrew science. Waiter's and double-lever corkscrews or camera and compounds eyes, on the other hand, display differences that apparently make differences to how each kind of corkscrew performs its cork-removing job.

In considering the differences between visual and rewired auditory cortex, we must ask whether they amount to differences that make a difference to how each performs its visual perception job. Making such an assessment more difficult than the comparable judgment in the corkscrew case is the fact that the visual and rewired audio cortexes do not perform the *same* visual perception job. Processes in visual cortex allow finer-grained visual discriminations than the processes in rewired auditory cortex permit. Accordingly, the question, "Do the visual and rewired auditory cortexes do the same thing in different ways?" becomes "*To the extent that* the visual and rewired auditory cortexes do the same thing, do they do it in different ways?"

We favor a negative answer to this question. In discussing the ways in which rewired auditory cortex "tries" to become visual cortex, we noted that it develops columns of orientation cells, 2D maps, and horizontal connections between regions of orientation-sensitive cells. However, just as the parts of a bicycle make for a poor clock, so the materials of auditory cortex face limitations in how well they can realize visual processing. To the extent that they do realize visual processing, they also resemble the structure of visual cortex. Correlatively, to the extent that their physical organization diverges from visual cortex, they fail to process visual information as successfully as visual cortex does. In a slogan, the sameness in function of visual and rewired auditory cortex owes to the sameness of their structure; the difference in function of the two cortexes traces to differences in their structure. Thus, they do not illustrate multiple realization. They do not do the same thing in different ways, but, insofar as they do the same thing at all, they do it in the same way. As candidates for evidence of multiple realization, the rewired ferrets fail both the "sameness" criterion and the "differently the same" criterion.

In this chapter we examined one sort of direct evidence for multiple realization, viz., neural plasticity. We argued that when measured against our official recipe for multiple realization, plasticity does not provide the existence proof of multiple realization that many philosophers have thought it does. But plasticity does not constitute the only form of direct evidence on offer for establishing multiple realization. In the next chapter we consider some different kinds of evidence—kinds that require close attention to the taxonomic practices of working scientists.

6

Evidence for Multiple Realization
Kind Splitting and Comparative Evidence

1. Memory and Kind Splitting
2. Cone Opsins and Trichromacy
3. Comparative Evidence for Multiple Realization:
   Octopuses and Birds
4. Direct and Indirect Evidence for Multiple Realization

1 Memory and Kind Splitting

In the previous chapter we examined whether there is direct evidence for multiple realization based on neural plasticity. In this chapter we look at other purported "existence proofs" that defenders of multiple realization favor.

That many cognitive capacities, such as perception and memory, exist in a variety of creatures may suggest the ubiquity of multiple realization. Consider the case of memory, so central to human cognitive life but frequently studied in model organisms ranging from mice to sea slugs. In everyday life we often speak generically of remembering facts, procedures, places, events, words, people, and so on. But psychologists distinguish many different kinds of memory. Endel Tulving, writing in 1972, remarked that "In a recent collection of essays on human memory . . . one can count references to some twenty-five or so categories of memory" (1972: 382). And then he proceeded to introduce an additional distinction! Are these different kinds of memory different realizers of the general state or process, *memory*? No. There is general consensus that the everyday notion of *memory* picks out a number of distinct and dissociable psychological processes. The psychological phenomenon of memory is disunified. Memory is not one psychological process with diverse "lower-level" psychological realizers. Rather "it" is a collection of related capacities and phenomena.[1] Declarative and non-declarative memory are not diverse realizers of memory because they fail to satisfy clause (i) of our recipe— they are not, according to psychologists themselves, genuinely the same phenomenon after all, and so differences in their realization do not constitute the multiple realization of a single capacity.

At this juncture it will be useful to remember what we have and have not claimed. We have begun to argue that establishing the fact of multiple realization is much more difficult than many philosophers and cognitive scientists have assumed it would be. But we have not argued for its impossibility. Nothing in principle would stop psychologists from classifying memory systems and processes in a way that satisfies our Official Recipe. But as a matter of fact they have not done so. Indeed, the study of memory contains many examples where neuroscientists introduce new distinctions among kinds of memory processes when they discover them to be implemented in different brain processes. In other words, memory researchers have tended to engage in taxonomic "kind splitting." And the reason is plain: Neural differences in memory systems typically manifest themselves in psychological differences. Most dramatically, memory systems are typically dissociable—components can be selectively damaged or interfered with while leaving the others more or less intact. A survey article from 1996 reports: "Current views recognize a number of different forms or aspects of learning and memory involving different neural systems in the brain . . . On the other hand it is likely that under normal conditions many or all of these brain–memory systems are engaged to some degree in most learning situations" (Thompson and Kim 1996: 13438). This state of affairs is illustrated in Figure 6.1.

Of course the discovery of some difference in the neural realizers of memory processes does not mandate *kind splitting*. But kind splitting in the cognitive sciences is permissible and that is what is important for our purposes. In particular, kind splitting seems to be accepted practice when

---

[1] Memory is a widely discussed example in relation to multiple realization. See, e.g., Craver (2004, 2007) and Funkhouser (2014).

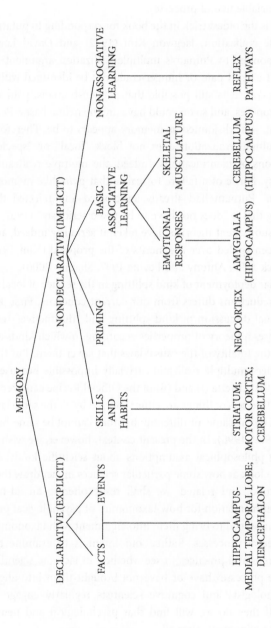

Figure 6.1 "A tentative taxonomy of long-term memory and associated brain structures" (Thompson and Kim 1996: 13439). Note that this figure does not include short-term memory systems; and further distinctions have been introduced since its time of publication. Reproduced with permission of American Association for the Advancement of Science.

what was thought to be a single psychological process is found to depend on multiple dissociable neural processes.

Kind splitting is the oldest trick in the book for responding to putative cases of multiple realization. Jaegwon Kim (1972) and David Lewis (1969) both responded to Putnam's multiple realization argument by pointing out that even if pain or hunger could not be identified with a single neural process, it is still possible that pain-in-humans, pain-in-dogs, pain-in octopuses, and so on could have unique neural bases. *Pain*, it might turn out, is as disjointed as memory appears to be. Therefore, they argued, multiple realization does not block "local" or "species-specific" reductions or identifications. Indeed, the multiple realization of pain is illusory, borne of a failure to realize that *pain*, like memory, should be "split." Putnam had already anticipated and rejected this response, writing that it "does not have to be taken seriously" (1967, in 1975: 437). But subsequent theorists have taken it seriously indeed, and much ink has been spilled over the merits of the proposal (Kim 1997, Fodor 1997, Block 1997, Antony and Levine 1997, Shapiro 2000).

But this familiar deployment of kind splitting in the service of local or species-specific reductions differs from our current concerns. First, the philosophers' usual discussion of kind splitting and "disjunctive" realization presupposes a theory of properties according to which kinds are individuated by the totality of the causal laws that cover them. But that assumption renders multiple realization trivially impossible because it does not appear that criteria (i) and (ii) of the Official Recipe can ever be jointly satisfied. Two sets of things are either covered by all the same laws or not. They can be the same, or different; but they cannot be same-but-different (Kim 1989, 1998). In the present context, however, we wish to avoid importing philosophical assumptions about scientific kinds; we choose instead to look at how some particular sciences in fact draw their taxonomies. Second, and related, we shall resist offering an ad-hoc proposal or recommendation for how taxonomies of psychological processes could be massaged to bring them into alignment with taxonomies of brain and neural processes. Rather, our intent is to examine the existing taxonomies and practices to see whether, in fact, the scientists who study those processes have or have not brought them into alignment. Do psychologists and cognitive scientists regularly engage in kind splitting? If they do we will find that psychological and neural taxonomies are in alignment, after all.

The answer in the case of memory is that kind splitting is both permissible and common. Carl Craver argues that neuroscientists are committed to what he calls the principle of *No Dissociable Realization*:

> (NDR)   Instances of a natural kind have one and only one realizer. If there are two distinct realizers for a putative instance of a kind, there are really two kinds, one for each realizer. (2004: 962)

Note that *No Dissociable Realization* depends on the would-be realizers being fully *dissociable*—that is, each can occur without the other. That is a much stronger requirement than merely being different, as our criterion (ii) requires. If two things are dissociable, there is no doubt that they differ. This explains Craver's claim that dissociations mandate kind splitting; but other sorts of differences between kinds might merely make kind splitting permissible rather than obligatory.

Craver derives the *No Dissociable Realization* principle from the example of the dissociation of declarative and procedural memory systems. As he tells the story:

> The argument for splitting typically opens with the story of H. M. .... In an effort to relieve life-threatening epileptic seizures, H. M. consented to experimental surgery removing bilaterally a structure known as the hippocampus. H. M. can still read and write, he can learn new skills, he can remember much of his childhood, and his IQ if anything increased after the surgery. The tragedy of H. M.'s life is that he permanently lost the ability to retain new memories for facts and events. He has lost declarative memory but has maintained procedural memory . . . From the case of H. M. and from studies in nonhuman animals, researchers have concluded that the mechanisms for procedural and declarative memory are distinct. Since the two are realized by distinct mechanisms, the kind "memory" is confused, lumping together what we now know to be two distinct kinds of memory. (Craver 2004: 961)

And Craver argues that the *No Dissociable Realization* is widely endorsed by psychological and brain sciences, citing both philosophers and psychologists. For example, Daniel Schacter writes:

> we have now come to believe that memory is not a single or unitary faculty of the mind, as was long assumed. Instead, it is composed of a variety of distinct and dissociable processes and systems. Each system depends on a particular constellation of networks in the brain that involve different neural structures, each of which plays a highly specialized role within the system. (Schacter 1996: 5)

As we've noted, if Craver is right then the discovery of dissociable realizers is a sufficient condition for kind splitting.

The existence of dissociable realizers is compatible with the existence of multiple realization, but does not provide evidence for it. The "pre-split" psychological kind (e.g., *memory*) exhibits impairment in the absence of any of the dissociable sub-systems that compose "it." The remaining sub-systems realize a similar but degraded process. And if we focus on the post-splitting systems, we see why this is the case. For they do not each redundantly do the same thing. Rather they each carry out different functions, sometimes overlapping but frequently not. This, after all, explains their dissociability and also the impairment of the "one" system that remains after the loss of a dissociable "realizer." Dissociable realizers are not examples of same-but-different, they are yet again examples of different-and-different.

2 Cone Opsins and Trichromacy

Ken Aizawa and Carl Gillett have argued that the example of kind splitting in memory is misleading, and that cognitive scientists do not generally split kinds when they discover diverse neural realizers (2011; Aizawa forthcoming). If correct, this suggests that memory might be multiply realized after all, for now the diversity of neural realizers that, we argued, implement a diversity of kinds of different memories might after all implement just a single kind, i.e. *memory*. Aizawa and Gillett use the example of variation in photosensitive proteins in human retinal cone cells to illustrate a case where cognitive scientists do not split kinds when they discover neural differences. According to Aizawa and Gillett, the cone cell example provides a better model for thinking about the practices of the cognitive and brain sciences, and one that supports multiple realization. Aizawa and Gillett agree with us that the question of multiple realization should be decided by evidence, and they bring to the table a detailed case. Yet they arrive at different conclusions than we do, so it will be worthwhile to examine their argument at length.

But before examining the details of Aizawa and Gillett's argument, we must emphasize that in our view cognitive scientists might sometimes take the kind-splitting route and sometimes not. Aizawa and Gillett's mobilization of the cone opsins example appears to be directed at an opponent of multiple realization who thinks that variation in putative realizers must always result in kind splitting. That is not the version of the kind-splitting argument that we have offered herein. We merely proposed that some prima facie examples of multiple realization fail because scientists split the commonsense or familiar kind, as in the example of memory. But we are happy to admit that scientists do not split kinds every time they discover a variation in would-be realizers. As critics of the species-specific strategy have long pointed out, splitting kinds for every variation in nature would result in micro-kinds lacking in generality and therefore unsuitable for scientific explanation (Horgan 1993b). We all agree that would be bad.

Why do Aizawa and Gillett take us to hold that kind splitting is always mandatory? The answer is that we do believe that *under certain assumptions* kind splitting would be mandatory. For example, suppose that all scientific kinds were individuated by the exceptionless causal laws in which they figure. Then kind splitting would be mandatory: If any causal difference were to be found among would-be realizers, the putative kind that they realized would have to be split.[2] However, we do not subscribe to this theory of kind individuation, and in fact view it as destructive to the interesting empirical question about the prevalence of multiple realization. As we've urged repeatedly, issues concerning multiple realization should not be decided a priori, as they would be if followed simply from one's theory of kind individuation.

We can now turn to the details of Aizawa and Gillett's argument. Normal human color vision is trichromatic, meaning that normally sighted human beings can match almost any color sample by mixing three different "primary" lights (Surridge et al. 2003). Trichromacy is "normal" for human beings both in the sense that the majority of the population is trichromatic and in that it appears to have been selected for by natural selection, although an adaptationist explanation for trichromacy remains controversial (the most commonly told adaptationist story connects mammalian trichromacy to improved foraging, but the details do not concern us here). Trichromacy and even more discriminating forms of spectral sensitivity in vision are common in fish, birds, and

---

[2] That is, after controlling for how fine-grained the causal laws must be. If a kind is individuated by coarse-grained causal laws and its realizers are individuated by fine-grained causal laws, then differences in the realizers might not require splitting the coarse-grained kind. But that will not be an example of multiple realization, either, for the fine-grained differences among the realizers will be merely individual differences in the coarse-grained realized kind. For more on grain mismatches, see Bechtel and Mundale (1999); also, Sachse and Esfeld (2007) and Soom et al. (2010).

invertebrates. In contrast, the majority of mammals lack trichromacy, the exceptions being many (but not all) primates.

Now consider the neurophysiology of the mammalian eye. In human beings and other primates, the retina contains two systems of photoreceptors. One shows sensitivity to low-intensity light stimulation and thus mainly operates in nighttime (nocturnal) conditions. This system receives inputs from a distinctive kind of photoreceptive cell in the retina—the rod—that is responsive to light intensity but mainly insensitive to spectral wavelength. The nocturnal system is therefore not considered to be a "color" vision system. The second photoreceptive system displays sensitivity to high-intensity light stimulation, and thus mainly operates in daytime (diurnal) conditions. This system receives input from three distinct retinal photoreceptive cell types, variously called the short (S), middle (M), and long (L) wavelength cones; or (somewhat misleadingly) the blue, green, and red cones. Because stimulation of the three cone types causes our color experiences, almost any color experience can be matched by a combination of three lights with distinct spectral compositions. The capacity for that matching behavior is the phenomenon of trichromacy.[3]

The different spectral receptivity of the different cone types results from differences in the photoreceptive protein pigments contained within the cells:

The absorbance spectra of the S-, M-, and L-cone photopigments overlap considerably, but have their wavelengths of maximum absorbance (λ_{max}) in different parts of the visual spectrum: ca. 420, 530, and 558nm, respectively. (Sharpe et al. 1999: 3)

The photopigments are opsins, proteins thought to have evolved only once, and around 700 million years ago: "Opsins are the universal photoreceptor molecules of all visual systems in the animal kingdom" (Shichida and Matsuyama 2009).

That differences in cone opsins explains the trichromacy of the human visual system is the starting point for Aizawa and Gillett's critique of kind splitting. For although vision scientists recognize three kinds of cones, they recognize many more than three kinds of opsins. Moreover, some of

---

[3] There are numerous good textbook explanations of color vision neuroanatomy. We are especially fond of Hardin (1988), Purves et al. (2001), and Purves and Lotto (2003).

the variations in opsins are quite common within the trichromatic human population:

A number of studies have documented the existence of polymorphisms in the green and red photopigments. For the red photopigment, it has been estimated that roughly 44% of the population has an amino acid chain, often designated Red (ala^{180}), that has an alanine at position 180, where about 56% of the population has an amino acid chain, often designated Red (ser^{180}), with a serine at position 180. For the green photopigment, it has been estimated that roughly 94% of the population has an amino acid chain, often designated Green (ala^{180}), that has an alanine at position 180, where about 6% of the population has an amino acid chain, often designated Green (ser^{180}), with a serine at position 180. These different amino acid chains contribute slightly different absorption spectra, which are properties that they contribute to the realization of normal human colour vision. For example . . . the wavelength of maximum absorption, λ_{max}, for Red (ala^{180}) is 552.4nm and that the Red (ser^{180}) λ_{max} [is] 556.7.10 Thus, one might expect that the property of having normal colour vision would be subtyped. (Aizawa and Gillett 2011: 212)

Several kinds of opsins appear in retinal cones. But cones are not sub-typed according to these variations. Aizawa and Gillett take this to show that the kind-splitting strategy is misguided, and consequently that critics of multiple realization cannot explain away prima facie evidence by appeal to kind splitting: "Actual practice with regard to normal colour vision does not follow the property splitting strategy. Instead, vision scientists appear to accept, or at least tolerate, the exist-ence of non-identical realizers of the higher level property of normal colour vision" (Aizawa and Gillett 2011: 213).

Moreover, Aizawa and Gillett encourage us to conclude that multiple realization is ubiquitous on the basis of the cone opsin case, for two reasons. First, the kind of variation illustrated in the case of cone opsins is typical in organic chemistry—most every protein in any neuron can be found in variations. Second, because their understanding of multiple realization presupposes that realization is a relation between properties of parts and properties of wholes, Aizawa and Gillett believe that vari-ation at any compositional level generates multiple realization at every higher compositional level.[4] Consequently, they say, "Insofar as there is

---

[4] Our approach to multiple realization, in contrast, does not make any assumptions about the right account of realization. As we explained in Chapter 2, we believe that realization is posited to explain how physicalism is compatible with widespread multiple realization, not vice versa.

realization and multiple realization of colour vision by the apparatus of the eye, there will be at least that much realization and multiple realization in the entirety of the visual system" (Aizawa and Gillett 2011: 208). This is important for Aizawa and Gillett because cone opsins are not the only causes of trichromacy, but opsins are parts of visual systems that exhibit trichromacy—or parts of parts of visual systems, and so on. So variation in opsins results in multiple realization all the way up the compositional hierarchy, according to Aizawa and Gillett.

Our response to Aizawa and Gillett begins with a cautionary note. To say that human beings are trichromats or that normal human color vision is trichromatic is to say that normal human beings exhibit a certain behavioral pattern. Trichromacy is the capacity to do a certain task—to match a sample using three primary lights. "Being trichromatic" is more like "being graceful" than it is like "being a vertebrate"; it is a behavior or effect that might have many causes. This makes us hesitant about whether the example of "normal human color vision" is an example of an internal or cognitive process at all, rather than the output of such a process. But as it happens, the trichromatic behavior has a specific and relatively localized cause, viz., having three photoreceptor types connected in the opponent-processing configuration that is normal for human beings. Someone who had three cone types but lacked the normal opponent processing neural circuitry would probably not exhibit trichromacy—at least, not normal human trichromacy. Certainly that is the case if normal opponent processing requires having a normal visual cortex. Cortical achromatopsics are entirely color blind, not merely color deficient. But their retinal cone systems are intact. So we shall take Aizawa and Gillett not to be claiming merely that the trichromatic behavior has multiple causes, but that the normal human visual system is multiply realized because there are variations in its parts.

We argued in Chapters 3–4 that not all variation within kinds should count as instances of multiple realization. But should variation of the sort present in cone cells count? As a first step toward answering this question, we must make it more specific. When we ask whether variations in cone opsins count as multiple realizations of normal human color vision, strictly we will understand that as a question of whether *human-opponent-process-visual-systems-with-ala^{180}-cone-retinas* and *human-opponent-process-visual-systems-with-ser^{180}-cone-retinas* are distinct realizers of the kind *human-opponent-process-visual-system*. Neither we nor Aizawa and Gillett believe that actual scientists use those kinds in their explanations. But this is a way of allowing our discussion to proceed to the question of which patterns of variation we find in the world, and it does so without requiring us to settle our philosophical disagreements in advance.

Aizawa and Gillett's example of variations among cone opsins differs in significant respects from Craver's discussion of memory. True, the variation in cone opsins makes a difference to the discriminative capacities of the people who have retinas of each sort. But vision scientists regard those differences as falling within the range of normal color vision. Thus, the differences do not look anything like the differences between declarative and procedural memory, which indicate a need to split memory into at least two kinds. Trichromats, despite variation in cone opsins, can all match a target light with three primary lights. Moreover, they are all trichromats for the same reason: They all have three kinds of cones. The differences between those with ala^{180} cones and those with ser^{180} cones are simply individual differences among normal human color perceivers.

The example of the variation in normal human color vision due to variation in cone opsins fits the first three quarters of our recipe for multiple realization:

(i)   There is a taxonomic system according to which trichromatic perceivers are all the same—it sorts its kinds by their trichromacy.

(ii) There is a taxonomic system that distinguishes among trichromats—it sorts them by their peak sensitivities, say.

(iii) The differences that explain the variations in peak sensitivity are relevant to trichromacy—the presence of three kinds of cones with three different photopigments.

But consider the last criterion that the variation in the opsins must meet if it is to qualify as multiple realization:

(iv) The differences among the opsins are distinct from individual differences among trichromats.

Differences in opsins fail this criterion, because the variation that these differences produce in the color vision capacities of human beings falls within what vision scientists regard as a normal range. It's the kind of variation that one might find in the cork-removing abilities of wood and metal waiter's corkscrews. On the other hand, it's nothing at all like the variation one finds between procedural and declarative memory.

In short, all trichromats with human S-cones, to continue the example, have S-cones insofar as their opsins are similar. Molecular differences among human S-opsins are reflected in differences among perceivers; and the similarity of the perceivers is explained by the similarity of their S-cones. They are not, in short, differently the same—different ways of doing the same thing. They are differently different and samely the same. The cones with ala^{180} and ser^{180} opsins do the same thing in the same way, and where they do different things, e.g. exhibiting slight differences in peak sensitivities, they differ biochemically. So this variation in human cone types does not turn out to be a case of multiple realization, after all.

At this point it is useful to remember that questions about multiple realization are always specific and contrastive. We do not ask: Is normal human color vision multiply realized? Nor do we ask: Are cones with ala^{180} and ser^{180} opsins multiple realizers? We ask: Is normal human color vision multiply realized by cones with ala^{180} and ser^{180} opsins? And we argued for a negative answer. Yet the negative answer to that specific and contrastive question is fully compatible with normal human color vision being multiply realized by some other difference in would-be realizers. That is, it is compatible with a positive answer to some other specific and contrastive question.

For the moment we claim only to have argued for two conclusions. First, that the example of variations in human cone opsins does not make for concrete direct evidence of actual multiple realization of a psychological capacity. Second, and more to the point, Aizawa and Gillett have not shown that the modest kind-splitting strategy we embrace is wrongheaded. We have observed that in some prima facie cases of multiple realization, scientists in fact decide to split kinds, thereby removing the appearance of multiple realization. Aizawa and Gillett point out that kind splitting is not mandatory, and we agree. But we further used their example of variation in human cone opsins to emphasize our contention that only some kinds of variation in the world count as multiple realization. We argued that variation in retinal S-cone opsins among normal trichromatic humans fails as an example of multiple realization because it is not the right kind of variation. Obviously memory and cone opsins are only two examples among multitudes, but we believe these examples are not anomalous. If such cases can be shown to be typical or expected, then this would provide indirect evidence that multiple realization is far rarer than philosophers have supposed.

As should by now be obvious, we do not claim that multiple realization never occurs. Whether the aims and methodologies of any particular pairs of sciences recommend bringing their taxonomies into alignment is likely to be highly variable, and may be a matter of dispute among scientists themselves. If the sciences operate largely independently of one another, that may result in mismatched taxonomies and multiply realized kinds. In the case of engineered kinds, we often invest a great deal of effort into designing multiply realizable things so that their parts can be repaired or replaced interchangeably. But psychological and neural systems were not designed to have interchangeable parts, and cognitive and brain scientists tend to work in relatively close cooperation. They share data, techniques, experimental designs, stimulus paradigms, and operationalized procedures. And they tend to split and combine kinds in order to keep their taxonomies in alignment when it is possible to do so. This coevolution of the mind and brain sciences (Churchland 1986) adds up to what Carl Craver describes as a "mosaic unity" (2007).

Let us consider some additional cases to see if they provide direct evidence for multiple realization.

3 Comparative Evidence for Multiple Realization: Octopuses and Birds

The examples we have considered so far—neural plasticity, memory, and cone opsins—purport to show that psychological processes are multiply realized within individual cognizers or among members of a species. But Putnam's original multiple realization argument sprang from the claim that one and the same psychological state or process, e.g., pain, is present in members of different species and even phyla—by both chordata and mollusca, by human beings and by octopuses. Perhaps the similarity within individuals over time and between individuals of a single species makes it too easy to explain away multiple realization. Maybe we need to look for more dramatic differences. Are there examples of multiple realization across species? Insofar as we have reason to believe that non-human animals undergo psychological states or processes, do their

				SENSORY					
	Temporal	Spatial	Punctate	Incisive	Constrictive	Traction	Thermal	Brightness	Dullness
1									Dull	1
2	Quivering		Pricking		Pinching			Smarting		2
3	Pounding	Shooting		Sharp		Tugging	Hot		Heavy	3
4			Stabbing	Lacerating	Crushing		Searing			4
5										5

			AFFECTIVE			EVALUATIVE	
	Tension	Autonomic	Fear	Punishment	Misc.	Anchor Words	
1						Mild	1
2	Tiring					Discomforting	2
3	Exhausting	Sickening	Frightful	Gruelling	Wretched	Distressing	3
4		Suffocating	Terrifying		Blinding	Horrible	4
5				Killing		Excruciating	5

Figure 6.2 Patient evaluations of pain sensations. Melzack and Torgerson 1971. Used by permission.

realizations in non-human animals differ from their realizations in human beings?

Consider the example of pain. Pain in human beings is very well studied, due in part to its obvious practical and clinical importance in our lives. Pain researchers distinguish numerous varieties of pain. And despite many philosophers' insistence on the ineffability of sensations, patients do quite well at qualitatively and quantitatively describing their pains (Figure 6.2).

Moreover, we have a fairly good understanding of the basic neural mechanisms that mediate pains. The pain system is a sensory system. Pain is normally triggered by the activation of specialized sensory neurons, the nociceptors, that detect stimuli of various sorts: Mechanical, chemical, thermal, etc. The specificity and intensity of the nociceptor activation cause some of the characteristics of pain experience. Other features of pain experience appear to be associated with the features of the connected neurons by which the nociceptors signal the spine and brain—the notorious C-fibers, Aβ-fibers, and Aδ-fibers, associated with "fast" and "slow" pains. All the way from nociceptors to the cortex we can match the features and structure of pain experiences with the features and structures of the pain sensory system in human beings (cf. Polger and Sufka 2006). Pain experience is also modulated by activity outside the pain sensory system; and notoriously pain can occur even when some parts of the system are missing, as with phantom pains. But many aspects of pain experience have specific neural bases and explanations.

Nevertheless, difficult questions about the realization of pain in different creatures arise. Nociceptors, like light-sensitive cones, are evolutionarily ancient and common across animal species. Nociceptors are present in mammals and mollusks; and also in fish, sea slugs, and fruit flies. The molecular predecessors of nociception are found in single-celled organisms that respond discriminately to noxious stimuli (Smith and Lewin 2009). All animals that experience pain do so in part because they have nociceptors. Octopuses and other mollusks have nociceptors, but nociception by itself is generally considered insufficient for experiencing pain. Moreover, octopuses lack spines, not to mention cortices. Doesn't that show that if octopuses have pains, their pains are not realized in the same way as mammalian pains?

Once again, we urge caution. If we have confidence that octopuses experience pains this must be because they exhibit various pain-like behaviors, and, e.g., because they have nervous systems that include nociceptors. Their behavior discriminates among noxious stimuli in just the ways that we would expect given the kinds of nociceptors that they have—for instance, octopuses lack thermal nociceptors and, unsurprisingly, show no aversion to cold; but they do have mechanoreceptors and, again unsurprisingly, display an aversion to damaging mechanical stimuli. Furthermore, octopuses attend to bodily damage in ways similar those that vertebrates exhibit, and that are not observed in other mollusks such as squid. This fact is thought to mark an important distinction between pain and mere nociception (Alupay et al. 2014). Thus, we think the right conclusion to draw about octopuses is that their pain experiences differ from human experiences in the ways that we would expect for creatures that have some parts of the familiar mammalian pain sensory system but lack others.

Although octopuses are large-brained creatures, they do not have a neocortex as mammals do, much less a neocortex of the human sort. That seems to be the kind of huge neuroanatomical difference that Putnam had in mind when he confidently claimed that humans and octopuses could not have brain states of the same sort. But recent neuroscience puts less stock in these kinds of gross anatomical differences than philosophers have expected. Consider again the case of memory. Mammalian memory crucially relies on processes in the hippocampus, a brain structure that octopuses lack. Yet octopuses exhibit memory behaviors, such as acquired aversions to stimuli. How can that be? The answer appears to be of the form we suggested in the case of pain:

the similarity in the architecture and physiological connectivity of the octopus MSF-VL system to the mammalian hippocampus and the extremely high number of small interneurons in its areas of learning and memory suggest the importance of a large number of units that independently, by *en passant* innervation, form a high redundancy of connections. As these features are found in both the octopus MSF-VL system and the hippocampus, it would appear that they are needed to create a large capacity for memory associations. (Hochner et al. 2006: 315)

So when it comes to memory, octopus brains appear similar to human brains after all. In the case of human beings and octopuses, these structures do not appear to have a common ancestor—although they involve molecular changes in neurons that did evolve in a common ancestor. Rather—and contrary to the expectations of Block and Fodor (1972)—convergent evolution produced similar realizers in human beings and octopuses: "a convergent evolutionary process has led to the selection of similar networks and synaptic plasticity in evolutionarily very remote species that evolved to similar behaviors and modes of life. These evolutionary considerations substantiate the importance of these cellular and morphological properties for neural systems that mediate complex forms of learning and memory" (Hochner et al. 2006: 315). We might not have expected it, but similar psychological processes are realized by similar cellular and neural structures in human beings and octopuses.

Now the capabilities of octopuses are unique among mollusks, even if we discount their purported abilities to predict the winners of World Cup matches. So it might be thought that octopus memory—like octopus eyes, discussed in Chapter 2—is another exception that proves the rule. Octopuses are special, one might think. Maybe octopuses evolved neural mechanisms for memory that are similar to those in human beings; but that is mere coincidence. It is much more likely for evolution to produce multiple realization than for it to produce similar realizers. But that is not the case. Over and over neuroscientists are discovering similar structures in diverse organisms.

Mammalian sensory systems typically involve specialized sensory cells (retinal rods and cones, nociceptors, and so on) that project into specialized regions of the neocortex. But those specialized cortical areas appear to be absent in non-mammalian species that lack a neocortex altogether. Despite that difference, creatures such as birds and fish can do many of the tasks that mammals can do. The question therefore arises, as

Harvey Karten (2013: R13) describes it, "How is the avian/reptilian brain organized? How is sensory information processed in a manner resulting in virtually identical outcomes to that of mammals, in the seeming absence of a neocortex?" Karten is describing a question that arose by the 1960s, at the same time that Putnam was arriving at his own hypotheses. But Karten arrived at a quite different conclusion from Putnam's. Describing his work on a brain structure of birds known as the dorsal ventricular ridge (DVR), Karten answers his own question:

Back in 1969, from analysis of their connections, cellular morphology, physiology and histochemical properties, I concluded that, in avian and reptilian brains, the major sensory cells and circuits directly comparable to those of the mammalian neocortex are found in the DVR and adjoining dorsomedial pallial "wulst". These cells performed similar, or even identical, computational operations to those of cortex. (2013: R13)

Taken at face value, Karten may seem to be describing the strong case for multiple realization: Similar or identical "computational operations" are performed by cortical cells and circuits in mammals and by DVR cells and circuits in birds. But that was not his conclusion at all. Instead, his 1969 speculation was that the similar cells and circuits in mammalian cortex and avian DVR are essentially the same structures doing the same things, and that they are evolutionarily homologous structures—they evolved from a common ancestor.[5]

In 1969 Karten's hypothesis was difficult to evaluate, and so he regarded it as a speculation. But contemporary molecular genetic techniques allow researchers to evaluate the evolutionary relatedness of the cells in the mammalian cortex and avian DVR by comparing genetic markers found in each kind of cell. Jennifer Dugas-Ford, Joanna Rowell, and Clifton Ragsdale tested Karten's hypothesis using these techniques, and concluded that the cells in avian DVR are indeed homologous to similarly organized cells in the mammalian cortex (2012). If this is correct, avian DVR is both analogous to and homologous with the mammalian cortex (Figure 6.3). As Karten puts it, "the avian brain contains cells and

---

[5] In evolutionary theory, homologous traits are those that evolve from a common ancestor. Neuroscientists sometimes use the term *homologous* for what evolutionary biologists call *analogous* traits—those that are currently similar. According to the research being discussed, avian DVR is homologous to the mammalian cortex in both senses: It is structurally and cellularly similar (analogous) and also comes from a common ancestor (homologous, in the evolutionary use of the term).

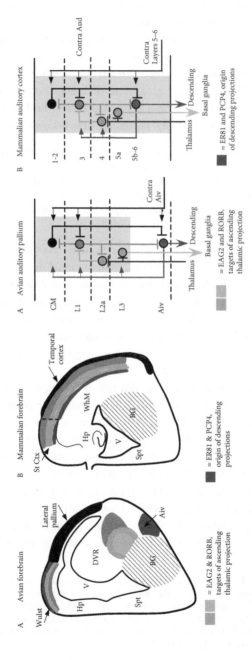

Figure 6.3 Left: Human auditory cortex circuits compared to avian DVR. Right: Macroanatomy showing locations of homologous neurons. Reprinted from Karten 2013, with permission from Elsevier.

circuitry which are nearly identical to those in the mammalian cortex, but disposed as nuclei rather than layers with interlaminar reciprocal connections" (2013: R15).[6] In our terms, Karten is saying that the differences in the spatial organization of avian and mammalian brains—nuclei versus layers—are not relevant differences. Human and bird sensory processes are not, after all, multiply realized. They are samely realized.

Similar subtleties can be illustrated using Brian Keeley's (2000) suggestion that the jamming avoidance response (JAR) in different electric fish—*Eigenmannia*, *Apteronotus*, and *Gymnarchus*—is multiply realized. These fish use a weak endogenously created electrical field to sense objects in their environments. But when two such fishes are close to one another, their electrical fields can interfere with one another, effectively "jamming" their electroreceptive senses. In order to avoid this, the fish must modulate the frequency of their fields to prevent interference, and the "cognitive task" facing the fish is "to decide whether to shift one's frequency up or down" (Keeley 2000: 452). Keeley argues that the JAR "is a clear and concrete example of multiple realization" (2000: 459) because there is a common algorithm used by the different fish to produce the response, but it is implemented in at least two relevantly different kinds of neurological systems (Figure 6.4; see also Heiligenberg 1991). Masashi Kawasaki puts the conclusion in the title of his paper: "Independently evolved jamming avoidance responses employ identical computational algorithms" (1993). So here appears to be a case that plainly satisfies (i)–(ii) in our four-ingredient recipe for multiple realization: There is a single cognitive process that is realized by at least two different physiological mechanisms.

But what about ingredients (iii) and (iv)? Are the differences between the realizers of the JAR relevant differences, and do they contribute to the

---

[6] Notice that the "cells and circuitry" that Karten proposes to identify across species here are not much like c-fibers, activated areas in functional magnetic resonance imaging pictures, or other more cartoonish candidates for brain process kinds. Similar observations apply to the circuitry discussed with respect to octopuses, above, and electric fish, below. Understanding the identity theory involves updating our expectations about brain processes themselves. (We are grateful to an anonymous reader for pressing us to make this explicit.)

We learned about Karten's hypothesis and research into the avian DVR from Michael Tye's talk at the *Conscious Thought and Thought About Consciousness* conference at the University of Mississippi. We do not think that Tye would embrace our interpretation of the evidence.

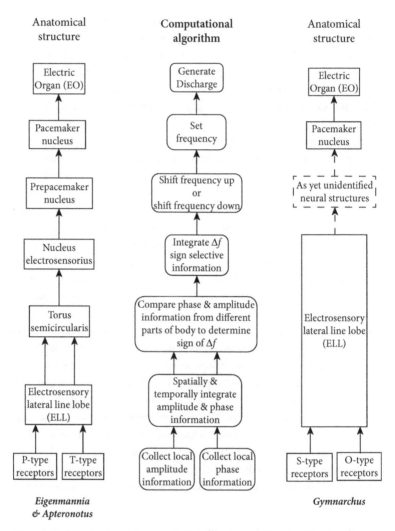

Figure 6.4 "Comparison of anatomical instantiations of Heiligenberg's jamming avoidance response algorithm in three genera of weakly electric fish" (reprinted from Keeley 2000: 456). Copyright 2000 University of Chicago Press, used by permission.

sameness of the JAR processing or only to species variations between the different fishes? It seems to us that the answer is none too clear. If we examine, for example, the time comparison circuits in the various species (Figure 6.5), we find the kinds of similarity that led Karten, in the case of

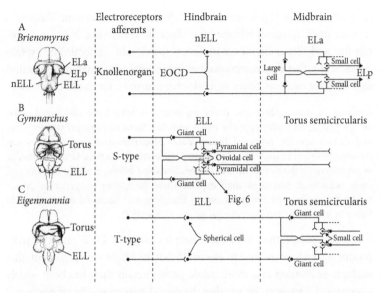

Figure 6.5 Time comparison circuits two wave type electric fish species (B and C) that produce the JAR response. The top species (A) is a pulse type that does not implement the JAR algorithm but instead avoids interference by making use of endogenously available signals caused by its own electrical discharges—electric organ corollary discharges (EOCD). Artwork reprinted from figure 5 of Kawasaki 2009. Used by permission of the Zoological Society of Japan.

the mammalian auditory cortex and avian DVR, to consider them to be basically the same. And Kawasaki writes, "[t]he organization of the time-comparison circuit is remarkably similar among *Brienomyrus*, *Gymnarchus*, and *Eigenmannia*" (2009: 593).

On the other hand, the time comparison circuits occur in different areas of the fish brains—hindbrain versus midbrain—and consist in physiologically distinguishable neuron types—pyramidal cells versus small cells on the output side (Figure 6.5, right side). Are these differences relevant differences? Do they contribute to the common JAR-producing process in both species, or only explain differences in the species behaviors? If the answer to questions like these is yes, then the JAR system looks like a good example of multiple realization. But, speaking of the time comparison circuits, Kawasaki comments that "The only differences are that neurons responding to the time difference appear only in the midbrain, and that midbrain 'sign selective' neurons are abundant in *Gymnarchus* but relatively scarce in *Eigenmannia*" (2009: 594). As we understand Kawasaki, it is an open question whether the observed differences between *Gymnarchus* and *Eigenmannia* are irrelevant (hindbrain, midbrain) or explain differences rather than commonalities (pyramidal cells, small cells; plentiful sign selective neurons, fewer sign selective neurons). Kawasaki concludes:

Analyses of time coding and decoding neural systems have uncovered some design principles underlying the behavioral functions of phylogenetically close and distant species. Some of these neural properties and performances are, however, difficult to understand in terms of behavioral function. Only knowledge of phylogenetic development and the relationships between neural circuits and their behavioral functions may explain these intriguing properties... Much remains to be learned about behavioral functions and neuronal mechanisms in fishes in important phylogenetic positions. (2009: 597)

From our perspective, both the examples of avian DVR and the JAR-producing system in weakly electric fishes highlight our argument that multiple realization is a more subtle phenomenon than has been widely appreciated. Determining whether the neural systems of *Eigenmannia* and *Gymnarchus* multiply realize the cognitive process that produces the JAR behavior is not settled simply by observing that they are different species, or that there are anatomical differences in their brains, or that they evolved the response independently. The questions will be whether the neural mechanisms that implement the process are relevantly different, and if their differences contribute to sameness rather than difference. The jury is out.

4 Direct and Indirect Evidence for Multiple Realization

Harvey Karten made his proposal about the similarity of neural structures between mammals and birds at just about the same time that Putnam made his proposal about the dissimilarity of the neural structures between human and non-human animals. Karten showed us his evidence. Putnam did not. In this chapter and the previous we have considered what Putnam's evidence might have been, and have examined some of the most compelling evidence that has been offered for actual multiple realization. Our conclusion is that multiple realization is much harder to find than Putnam expected, and much harder to find than current advocates of multiple realization usually assume.

Some advocates of multiple realization might deny the significance of our conclusion. As we noted earlier, Fodor claims, "The functionalist would not be disturbed if brain events turn out to be the only things with the functional properties that define mental states. Indeed, most functionalists fully expect it will turn out that way" (1981: 119). Why would functionalists "fully expect it will turn out that way"? Isn't that contrary to the spirit of multiple realization—wasn't the whole point that minds might have entirely different realizers in different creatures?

For Fodor, functionalists should not be surprised if all terrestrial realizers of minds are neural—even all actual mind realizers in the universe—because functionalism entails the multiple realiz*ability* of mental states and processes. According to Fodor we already know that minds are functional things—things defined by their inputs and outputs and abstractly characterized internal relations. Comparing minds to Coke machines, he says:

Any system whose states bore the proper relations to inputs, outputs and other states could be one of these [Coke] machines. No doubt it is reasonable to expect such a system to be constructed out of such things as wheels, levers and diodes . . . Similarly, it is reasonable to expect that our minds may prove to be neurophysiological . . . Nevertheless, the software description of a Coke machine does not logically require wheels, levers and diodes for its concrete realization. By the same token, the software description of the mind does not logically require neurons. As far as functionalism is concerned a Coke machine with states S1 and S2 could be made of ectoplasm, if there is such stuff and if its states have the right causal properties. Functionalism allows for the possibility of disembodied Coke machines in exactly the same way and to the same extent that it allows for the possibility of disembodied minds. (Fodor 1981: 119)

Because Fodor supposes that we already know that minds are functional things, we know that in principle they can be implemented by any suitably organized stuff: "In the functionalist view the psychology of a system depends not on the stuff it is made of (living cells, metal or spiritual energy) but on how the stuff is put together" (1981: 114). That is the nature of functional kinds. We can therefore be confident that minds are multiply realizable.

This confidence, moreover, springs from Fodor's belief that the functionalist framework has contributed to "major achievements of recent cognitive science" (1981: 122). Now we doubt that a commitment to functionalism has been essential to the achievements of cognitive science. But more importantly, we doubt that Fodor's defense of functionalism does not ultimately depend on considerations of multiple realizability. We shall return to each of these doubts. But for the moment the important thing to notice is that if we follow Fodor then we abandon the search for direct evidence of multiple realization. Instead, having renounced a need to find actual cases of multiple realization, he forces us to consider indirect reasons to believe that psychological states and processes are multiply realizable. And so a new chapter is in order.

7

The Likelihood of Multiple Realizability

1 Evidence for Multiple Realization

In the preceding two chapters we argued that many prima facie examples of multiple realization of the psychological by the neural do not withstand scrutiny. The simple reason is that a very distinctive sort of variation in the world must occur in order to fulfill the various roles attributed to multiple realization, and many interesting and surprising kinds of variation in nature lack these features. Real-world cases of neural plasticity that might initially seem to satisfy the requirements for multiple realization fail either because the observed neural differences result in *different* psychological capacities, or because the neural differences do not constitute *relevant* differences. Furthermore, claims to find multiple realization in kinds like memory and pain, on closer examination, neglect the taxonomic practice of cognitive scientists who often split psychological kinds along neurophysiological lines.

In short, many philosophers who have thought that examples of multiple realization are easy to come by have underestimated the resourcefulness of scientists—and thus the resources of scientific explanations and theories—for finding non-obvious similarities in nature. Consequently, we argued, gathering direct evidence for multiple realization is often more difficult than has usually been supposed. And, in concert with that observation, it is frequently possible to find interesting similarities among would-be realizers, after all—similarities that play a substantial role in explanations and theories.

But advocates of multiple realization have more in their arsenal than lists of purported examples. Some, recall, have argued that even if actual examples of multiple realization are few, nevertheless we should *expect* to find examples, and this expectation is enough to establish that psychological states are multiply realizable. This line of reasoning skips the step of locating actual examples of multiple realization and opts instead to argue from other considerations that multiple realization, rather than identity, is the more likely of the two hypotheses about the mind-brain relationship. As we noted previously, Ned Block and Jerry Fodor write, "the argument against physicalism [viz., the identity theory] rests upon the *empirical likelihood* that creatures of different composition and structure, which are in no interesting sense in identical physical states, can nevertheless be in identical psychological states; hence that types of psychological states are not in correspondence with types of physical states" (1972: 160, emphasis added). In this chapter we examine these other considerations in order to assess whether they provide the support for multiple realizability that proponents tout.

2 Evidence and Likelihood

Granting our suspicions about the "actual" cases of multiple realization, proponents of multiple realization might try to defend their position on the basis of more theoretical considerations. Moreover, they could reasonably contend, such an approach should not be regarded as a retreat or concession. Scientists often find reasons to accept a hypothesis that do not require direct examination of the phenomenon being hypothesized. Thus, for instance, Darwin never saw the evolution of a species through natural selection, but this did not prevent him from marshaling a number of convincing reasons for accepting that species do in fact evolve by natural selection. Similarly, even if cases of multiple realization have not been directly observed, perhaps its defenders can enumerate indirect reasons that should persuade us of its prevalence. Having done so, we might then wonder why we have not observed very many actual cases; maybe we have not looked hard enough or in the right places. But, it might be suggested, the promise of widespread multiple realization is all that must be demonstrated for realizationists to justify their conception of the relationship between psychology and neuroscience.

Arguments like Darwin's and those we present below can be characterized in outline as comparing two hypotheses, H1 and H2, and some observations, O:

L1.   Given the truth of H1, O is exactly what you'd expect.
L2.   Given the truth of H2, O comes as a surprise.
L3.   Therefore, H1 is to be preferred to H2.

The hypotheses that Darwin pitted against each other were his own theory of evolution by natural selection and the biblical view of species' origins. Darwin argued that many observations, such as those of extinctions, fossils, vestigial traits, and maladapted traits, were to be expected if species evolved by natural selection but were very puzzling if the biblical account of creation were true. On these grounds, Darwin justified the conclusion that natural selection was the likelier of the two hypotheses.

In the present context, we aim to examine how the hypothesis of multiple realizability stacks up against the hypothesis of identity, relative to various observations. That is, we intend to examine arguments of the same form as Darwin's but concerning different hypotheses and different observations. But before we fill in the content of these arguments, a few words about the style of argumentation that we're now considering are necessary.

The first point to note is that arguments like Darwin's—what philosophers of science today would call likelihood arguments—impose a direction of fit on the hypotheses and observations at issue. To see this, consider two ways that observations and hypotheses might relate to each other. On the one hand, we might ask: *Given some observation*, how probable is the truth of some theory? On the other, we might ask: *Given some hypothesis*, how probable is the evidence that we observe? To see the difference in these two questions, consider a simple hypothesis. Walking through the woods one day, you catch a glimpse of a large, hairy animal just as it ducks behind a tree. Your hypothesis is that you have spotted Bigfoot. We can now ask two questions:

B: What is the probability that Bigfoot is behind the tree given your observation of a large, hairy animal?

O: What is the probability that you observed a large, hairy animal given that Bigfoot just ducked behind the tree?

The answer to question B is this: The probability that Bigfoot is behind the tree is very small even given what you observed. Moreover, we can grant this even if the probability that Bigfoot lurks behind the tree is larger, given what you saw, than it would be without such evidence. We estimate that the probability is small because we think we know independently that Bigfoot does not exist. Chances are it was something other than Bigfoot that you observed darting behind the tree.

In contrast, the answer to question O is this: The probability that you observed a large, hairy animal given that Bigfoot just ducked behind the tree is extremely high. This is because, *if Bigfoot had just ducked behind a tree* while you were watching, then almost certainly you would have witnessed a large, hairy animal. The theory leads us to expect precisely what you just observed.[1]

The distinction above, between the probability of a hypothesis given an observation, which is often quite low, and the probability of an observation given a hypothesis, which may nevertheless be quite high, illustrates the narrow concern of likelihood arguments. From such reasoning, one cannot draw conclusions about the absolute probability of a given hypothesis. But one can still make judgments about whether an observation gives more support to one hypothesis than to another.

But this point about support must also be clarified. One might wonder, in the Bigfoot case, how much support does the observation of a large, hairy animal ducking behind a tree provide to the hypothesis? Or, in Darwin's case, how much support does the observation of vestigial traits provide to the theory of natural selection? It is a nice feature of likelihood arguments that one needn't try to answer questions like these in an absolute way, for such arguments always involve a comparative judgment between competing hypotheses. Darwin didn't have to assign a particular value to the amount of support that the presence of vestigial traits gave to his theory; he needed only to argue that the observation of vestigial traits

---

[1] Philosophers of science have recognized for a long time that hypotheses and theories do not entail observations all on their own (Duhem 1914), but only in conjunction with various assumptions. Thus, for instance, one must assume that Bigfoot is hairy if the present observation is to lend any support to the hypothesis. Similarly, Darwin assumed that God would not endow organisms with useless traits.

was *more* probable given his theory than they were given the biblical theory. Likewise, the observation of a large, hairy animal ducking behind a tree might be *more* probable given Bigfoot's existence than given a theory that says no big, hairy animals exist in the woods.[2] For present purposes, this feature of likelihood arguments—that they supply comparative rather than absolute judgments—is very welcome. We wish to evaluate whether multiple realizability does a better or worse job than our identity theory in accounting for various observations. To hope for something more, e.g. a claim about *exactly* how probable the competing hypotheses make some observation, is likely to hope for too much.

Finally and crucially, the use of likelihood reasoning requires that one take great care not to illicitly describe the contested observations in a manner that prejudges the truth of any of the hypotheses under consideration. The observation statement "A large, hairy animal just moved behind the tree" offers support to the Bigfoot hypothesis because, given what we believe about Bigfoot, we'd expect to see a large, hairy animal if Bigfoot really did live in the woods. However, the observation statement "Bigfoot just moved behind the tree," is ineligible as a description of the evidence for the existence of Bigfoot, for it assumes the truth of the hypothesis for which it's being offered as support. This constraint on observation statements—that they should not assume the truth of the hypothesis on which they bear—may seem too obvious to need mention; but as we'll soon see, it's sometimes violated.

We believe that realization theorists have employed likelihood arguments, even if they never claimed to do so explicitly. And so we arrive at Block and Fodor's likelihood arguments for multiple realizability.[3] They propose three kinds of evidence that they regard as consistent with—or to be expected given—multiple realizability, but to come as a surprise were the identity theory true:

(1) "The Lashleyan doctrine of neurological equipotentiality" (1972: 160).

---

[2] On the other hand, one might consider a hypothesis that makes the observations equal to or more probable than the Bigfoot hypothesis, e.g. the hypothesis that says a team of people dressed as Bigfoot are running loose in the woods. Hypothesis choice makes all the difference in likelihood arguments.

[3] While it might be anachronistic to assume that Block and Fodor had the current and technical sense of likelihood in mind, we believe that attributing to them a likelihood approach makes the best sense of the arguments they go on to provide.

(2) The "Darwinian doctrine of convergence applies to the phylogeny of psychology" (1972: 161).

(3) The "conceptual possibility" of machines psychologically similar to organisms "but physiologically sufficiently different from the organism that the psychophysical correlation does not hold for the machine" (1972: 161).

When evaluating these three lines of evidence, we must ask exactly how they are supposed to lend support to the hypothesis of possible multiple realization, i.e., multiple realizability. What do equipotentiality, evolutionary convergence, and the possibility of machine minds say about the relative likelihoods of multiple realization and mind-brain identity? Why should we expect these kinds of evidence if multiple realizability were true? And why should they come as a surprise given the truth of the identity theory?

3 Equipotentiality

As Block and Fodor present the neuroscientist Karl Lashley's idea of equipotentiality, it amounts to the claim that "a wide variety of psychological functions can be served by any of a wide variety of brain structures ... (For example, it is widely known that trauma can lead to the establishment of linguistic functions in the right hemisphere of right-handed subjects)" in contrast with the usual lateralization of linguistic functions in the left hemisphere of right-handed subjects (1972: 161). Lashley's theory of equipotentiality developed from his observation of rats that had been trained to run a maze. Following their training, portions of the rats' brains were surgically lesioned, and Lashley found that it was the amount of damage, regardless of the location of the damage, that affected the rats' post-operative performance on the maze (1929).[4] He concluded that the brain was made of "equipotential" tissue, i.e. tissue with the potential for performing the same functions without respect to its location in the brain.

Today, of course, neuroscience leaves no doubt that the doctrine of equipotentiality is false. The cortex is not like a uniform lump of clay. Far

---

[4] Outside of the ability to navigate the maze, Lashley reported that the lesions caused "many other disturbances of behavior, which cannot be stated quantitatively but which give a picture of general inadequacy of adaptive behavior" (1929: 176).

from consisting in equipotential tissue, the brain appears to be divided into very specialized areas, with numerous distinct kinds of cells. Controversy continues over whether the brain is organized modularly (Fodor 1983, Samuels 1998), but no one doubts that different areas of the brain typically perform distinctive functions, and that they often do so in part because they differ in the kinds of cells of which they are composed. Indeed there was already good evidence against equipotentiality in 1972. So one response to Block and Fodor would simply be to wonder why they take a false doctrine as good evidence for multiple realization. Indeed, because the doctrine is false, if multiple realization did make the observation of equipotentiality probable then this would be evidence against the hypothesis!

But presumably Block and Fodor did not wish to endorse equipotentiality as Lashley described it. For one thing, the Lashleyan doctrine appears to be the converse of multiple realization. Whereas the recipe for multiple realization requires sameness of psychological kind with difference in neurological kinds, Lashley (1929) believed that neural tissue is uniform throughout much of the brain. Rather than psychological kinds being multiply realizable in different kinds of neural substrates, the opposite, or so Lashley thought, was true: The same kind of neural substrate could realize different psychological kinds. Although equipotentiality was too ambitious a hypothesis, there is good reason to think that neural reuse is common—brain areas and processes are frequently multifunctional or pluripotent (Anderson 2010, 2014; Klein 2012; Figdor 2010; McCaffrey 2015). Yet we earlier argued that neural reuse does not fit the pattern of multiple realizability. So that cannot be what Block and Fodor had in mind.

If Block and Fodor did not mean to put much stock in Lashley's particular thesis of equipotentiality, why do they mention it? They probably saw in Lashley's work the kernel of a more familiar observation about brain function: Its plasticity.

In Chapter 5 we examined whether examples of neural plasticity provide direct evidence of actual multiple realization. We looked at a case of neural plasticity involving the brains of ferrets that had been rewired so that their auditory cortices received visual inputs. We argued that cases such as this one do not provide examples of actual multiple realization. To the extent that rewired auditory cortices and normal visual cortices perform the same function—to the extent that they result in the same capacities for visual discriminations—it seems as if they are physically similar too. And, moreover, differences in the visual capacities of the rewired and normal ferrets can be explained by differences in the organization of the rewired auditory cortex and the normal visual cortex. Hence, we concluded, this case of plasticity fails to satisfy the criteria for multiple realization. The visual abilities of the rewired and normal ferrets are not differently the same—they owe their similarity in function to their similarity in neural structure—and they are not the same but different—they differ in their visual capacities as a result of differences in neural structure.

But there remains another way to understand how plasticity might bear on the question of multiple realization that needn't commit to showing that actual cases of plasticity satisfy the criteria for multiple realization. The idea is that the familiar kinds of plasticity that we observe have not yet provided a clear example of actual multiple realization, but the prevalence of the phenomenon of plasticity is to be expected given the hypothesis of multiple realizability but surprising given the mind-brain identity hypothesis.

According to the Multiple Realizability hypothesis, similar psychological kinds can almost always be realized by different kinds of physical structures. Its competitor, the Identity Theory hypothesis, holds that there are important mind-brain kind identifications. The truth of Multiple Realizability is supposed to be incompatible with Identity Theory. Now we can construct an argument for Multiple Realizability that looks to plasticity as an indirect source of evidence:

(P1)   Given Multiple Realizability, plasticity is what one would expect to observe.

(P2)   Given Identity Theory, it would be surprising to observe plasticity.

(P3)   Thus, Multiple Realizability has a higher likelihood than Identity Theory.

On its surface, we find this plasticity argument plausible. It does seem that the truth of Identity Theory, in contrast to Multiple Realizability, would make the observation of ubiquitous plasticity surprising. We should not expect to observe much plasticity if sameness of psychological capacity usually entailed sameness of physical structure; and on the other hand, we should not be at all surprised to observe plasticity if Multiple

Realizability were true. Plasticity thus appears to give us good reason to favor Multiple Realizability over Identity Theory.

Digging below the surface, however, reveals a significant problem with this argument for Multiple Realizability. In our earlier discussion of the relationship between evidence and hypotheses, we noted that observation statements must not be presented in a way that prejudges the truth of the hypothesis for which they are offered as support. For instance, when wondering whether it was Bigfoot you saw ducking behind a tree, it would be inappropriate to consider an observation statement that asserted "Bigfoot just ducked behind the tree," for such a statement contains in its content an affirmation of the very hypothesis under investigation—that Bigfoot and not something else moved behind the tree. In contrast, the observation statement "A large hairy object just ducked behind the tree," is a permissible observation statement, because it remains neutral on the question whether Bigfoot or something else was observed.

All of this is obvious, of course. But the point is worth emphasizing because the argument involving plasticity makes just the same error as the one involving the observation statement that contains reference to Bigfoot. Plasticity shows that similar psychological capacities can be realized in different brain areas. But in what sense do these brain areas differ? Pursuit of this question returns us to the discussions of the rewired ferret from earlier chapters. Either the brain areas differ in a way that makes a difference to how the similar psychological capacities are realized or they do not. That is, they are differences of the waiter's and double-lever corkscrew variety, or they are differences of the red and blue waiter's corkscrews variety. Only relevant differences are of interest when hoping to establish multiple realization. Accordingly, if plasticity is to provide the kind of evidence that will help in the choice between Multiple Realizability and Identity Theory, it must assert something more like the following: Similar psychological capacities are realized in relevantly different brain areas within an organism following damage to the brain (or practice), *and these different brain regions bring about these psychological capacities in different ways.* Now if the characterization of plasticity already includes that plastic differences are relevant differences that make a difference, then of course one should expect plasticity given Multiple Realization, because plasticity *just is* the claim that Multiple Realization is correct. But for that very reason plasticity can't be evidence for Multiple Realizability. Moreover, we argued in Chapter 5 that we should not be sanguine about plasticity thus understood, for that interpretation of actual observations tends to neglect the nuances of relevant differences and relevant samenesses.

How might this problem be avoided? We need to articulate the observation of plasticity in a fashion that doesn't presuppose that the differences between normal and damaged brains are also differences that account for the similarity in the psychological capacities of the two brains. There are at least two ways to do this. One is to remove the problematic assumptions from plasticity, and introduce the thesis of *plasticity** as merely a variety of variation in brains due to use or damage. That's fine. But of course the claim that plasticity* is more probable given Multiple Realizability than the Identity Theory is then false. After all, the Identity Theory would predict changes in brains due to use or damage: It just denies that those changes qualify as relevant differences that count as different ways of doing the same thing. That was the point illustrated in Chapter 5 by example of the rewired ferrets, among other examples.

An alternative suggestion would be to understand plasticity in the assumption-laden way, but to argue that the assumptions are justified by other considerations. Perhaps physiological evidence suggests that some parts of the brain can never or only very rarely adopt the physically relevant properties of another part. In this context, "physically relevant" means something like "the physical properties of the neural structure in virtue of which the neural structure realizes the psychological capacity that it does." For example, in our examination of the ferret's rewired auditory cortex, we saw that areas of the auditory cortex can, perhaps surprisingly, organize into some of the physical structures that enable the visual cortex to perform visual processing, such as columns of orientation-sensitive cells. For that reason, we argued, the rewired ferrets do not supply the incontrovertible case for multiple realization that a less cautious examination might lead one to expect. But what if neuroscientists could establish that the rewired ferrets could see normally even though the physical properties by which the visual cortex realizes visual discriminatory capacities *could not possibly* be replicated in auditory cortex? (We hasten to remind you that this is not what the evidence shows.) That is, suppose it were shown that, by whatever means the auditory cortex replicated the function of the visual cortex, it must do so via relevantly different physical structures than those present in visual cortex. If so, we would have strong reasons to conclude that the different brain regions achieve their functions in different ways. Were that the case, the brain would not just be plastic, but *superplastic*—brain regions could perform many tasks without any underlying neural changes. Such a finding, along with evidence that the visual capacities of the rewired and normal ferrets were indeed quite similar, would make Multiple Realizability much harder to resist. Superplasticity—conceptualized so as to focus on the impossibility (or even the significant improbability) of one brain area replicating the relevant neural properties of another area, while at the same time realizing the same psychological capacity as that area—could then figure into the sort of likelihood argument we proposed above. Superplasticity would indeed be more probable given Multiple Realizability than given Identity Theory, because it violates the identity theorist's assumption that similar psychological function requires similarity in neural properties.

But then the evidential weight will be carried by the deeper neuroscientific theory of superplasticity that we are imagining, rather than on the observation of ordinary plasticity. Ordinary plasticity is no more evidence for superplasticity than is a man evidence for Superman. So we don't think that "Lashleyan" considerations of plasticity give any reason to think that Multiple Realizability is more probable than Identity Theory.

Of course no one has seriously proposed that brains or neurons are superplastic—e.g., that the structure of the brain or brain areas place no constraints at all on the functions that they can perform. But some neuroscientists endorse a related hypothesis, *neural degeneracy*; and some philosophers believe that degeneracy can be a reason to expect multiple realizability (Figdor 2010). Degeneracy is "the ability of elements that are structurally different to perform the same function or yield the same output" (Edelman and Gally 2001: 13763; cf. Tononi et al. 1999). Put this way, degeneracy involves the ability of different causes to produce similar effects; and we have been careful to point out that we do not deny that there are typically many ways to produce an effect. But the idea has also been applied in the neurosciences in a way that is more relevant to the question of multiple realization. Carrie Figdor writes, "In cognitive neuroanatomy, degeneracy is the claim that, for a given cognitive function F, there is more than one nonisomorphic (nonidentical) structural element that can subserve F, either within an individual at a time, across individuals, or within an individual across times" (2010: 428).

The reasons for hypothesizing widespread neural degeneracy trace back to the difficulty of localizing cognitive functions to particular brain areas or structures using either lesion or imaging studies (Figdor 2010: 428–9). The problem is not, as Lashley would have predicted, that there is no neural localization or specialization for cognitive functions; rather, the problem is that studies find different localization in different individuals and at different times (e.g., Price and Friston 2002, 2005; Noppeney et al. 2004). Summarizing the neuroscientific work, Figdor explains, "[t]he degeneracy hypothesis can explain these puzzling processing differences and data showing undeniable cross-subject specialization of function in cortex that have made Lashley's hypotheses untenable" (2010: 431). So, degeneracy is invoked to explain why the same functions and processes can be localized to different neural areas or systems in different individuals, or even in the same individual over time: There are multiple areas or systems that can produce the same results.

Now Figdor is aware of the kinds of concerns about empirical evidence for multiple realization that we have been pressing, and she does not argue that every example of degeneracy is an example of multiple realization. She allows, for example, that "duplicate anatomical areas subserving the same function, whether or not they function redundantly, would no more count as [multiple realizations] than the kidneys, which are both anatomically and functionally redundant" (2010: 433). Figdor also notes that some neuroscientists who hypothesize widespread degeneracy do not distinguish relevant from irrelevant differences, supposing that "a single difference in connectivity suffices for a distinct structure" (2010: 425). At best, some kinds of degeneracy would count as examples of multiple realization.

But, more importantly for our purposes, in defending multiple realization Figdor argues for a thesis that is weaker than the hypothesis that we call Multiple Realization, and that is compatible with our Identity Theory hypothesis. For Figdor argues that on the general hypothesis of degeneracy:

it is an open empirical possibility that when two neural structures count as the same by one criterion, they may count as different based on other criteria. Which of these physical differences will triumph in individuation in particular cases of criterial conflict cannot be determined a priori...In short, at least some degenerate systems are very likely to count as cases of [multiple realization], however the issue of realizer individuation is settled. (2010: 436)[5]

That is, some cognitive functions or processes may be multiply realized, and others not. But this much we have admitted; and it is entirely compatible with our Identity Theory, which advocates for important and explanatory mind-brain kind identities. In short, Figdor's opponent is a critic who outright denies the empirical possibility of multiple realization. That is not our view.

In the present context we are asking whether any general principles of neuroscience—equipotentiality, plasticity, superplasticity, degeneracy— would make it likely that cognitive states are multiply realizable even apart from actual observations of the kinds of variation that fit our Official Recipe. In particular, we are asking whether any general reasons exist to doubt the existence of important and interesting mind-brain identities. Figdor does not argue that degeneracy has that consequence, only that degeneracy makes multiple realization a reasonable hypothesis for a variety of cognitive functions. Moreover, Figdor agrees with us that for any specific cognitive function the question of multiple realization will only be resolved by actual evidence and the actual practices of specific sciences. She does not suggest that the issue can be resolved by considering possible but non-actual cognitive systems, nor solely by appeal to general principles like degeneracy.

In presenting and amending Block and Fodor's original claim that Lashley's doctrine of equipotentiality is to be expected given multiple realizability, our path has led to a comfortable resting point for skeptics (such as ourselves) about the prevalence of mind-brain multiple realiza- tion. As we have stressed on many occasions, the truth of the multiple realization thesis stands or falls on empirical considerations. Thinking about the role of plasticity in arguments for multiple realization points to one such empirical consideration. Proponents of multiple realization must focus their sights on how the brain implements psychological

---

[5] The question of "realizer individuation" mentioned by Figdor is the same that we discussed when we considered kind-splitting strategies, and is sometimes discussed under the guide of cognitive ontology revision. For philosophical discussion, see Klein (2012) and McCaffrey (2015). In our view these issues are also closely related to questions about the validity of neuroscientific experiments and protocols and their relation to cognitive and psychological experiments and explanations (Sullivan 2009, 2010, 2016).

processes. Support for multiple realization does not come from plasticity of any sort, but of only a special sort, e.g., superplasticity. It must be shown not only that different brain areas are capable of producing the same psychological functions, but that they do so without instantiating the relevantly same physical properties. The evidence we reviewed in the previous chapter does not support that prediction. Nor, have we argued, does consideration of neural degeneracy. The burden of establishing superplasticity is much greater than that of establishing mere plasticity.

4 Convergent Evolution

The second consideration that Block and Fodor (1972) invoke in favor of multiple realization is the prevalence of convergent evolution. Convergence, as the name suggests, occurs when different species evolve a similar trait subsequent to their divergence from a common ancestor. The torpedo-like shape of sharks and dolphins, or the wings of birds and bats, evolved only after each of these pairs of organisms diverged from their common ancestors. The same is true of the camera eyes of mammals and cephalopods, which display convergence in a remarkable number of features (Figure 3.1). Thus, the similarity in those traits cannot be explained by descent from a common ancestor. Rather, the organisms converged on a similar trait design, perhaps as a result of similar selection pressures. Rapid movement through water, for instance, favors a streamlined body shape, and that is likely why dolphins and sharks have similar body shapes. On the other hand, the opposable thumb that human beings share with other apes is not an example of convergence, because the common ancestor we share with other apes had an opposable thumb. This similarity between other apes and humans is explained by appeal to common ancestry.

What has all this to do with multiple realization? Block and Fodor claim that "[p]sychological similarities across species may often reflect convergent environmental selection rather than underlying physiological similarities" (1972: 161). As an example, they speculate that distinct lineages of organisms may have converged on the ability to experience pain. This possibility, they believe, is evidence for multiple realization because insofar as the lineages are distinct, we should expect physiological differences between them. Hence, two organisms may share, as a result of convergent evolution, similar psychological states while differing physiologically. This is just what we should expect if multiple realizability were possible; but not if minds and brains were identical.

Block and Fodor are not the only ones to think that evolutionary considerations favor Multiple Realization. Edelman and Gally invoke evolution in their argument for degeneracy, discussed above, for example (2001). But appealing to convergence as a source of support for multiple realizability raises a number of questions. A useful approach to the first question involves further reflection on the torpedo shape of sharks and dolphins. Speaking loosely, we might say that natural selection "cared" about designing a body shape that moved efficiently through the water, and this is why sharks and dolphins have a similar shape. However, natural selection did not "care" about how this body shape was constructed from the "building materials" that shark and dolphin bodies provided. Thus, the dolphin's shape comes about through an arrangement of a bony ribcage whereas the shark, which lacks bones and (more significantly) has no ribcage, owes its shape to a tough fabric-like arrangement of collagen in its skin.[6] The streamlined exterior shape of the body, as far as natural selection is concerned, is what matters. The internal structures that produce that shape are incidental, and that's why dolphins can do it one way and sharks another.

The analogous distinction in the case of minds is between "external" behaviors and the "internal" psychological processes that cause them. Remember that multiple realization is part of a realist view of minds; there really are internal psychological processes among the causes of behaviors. With this distinction in hand, we can now raise a difficulty with Block and Fodor's claim that the prevalence of convergence offers support for multiple realization. In this, they appear to follow Putnam's earlier suggestion that hunger in an octopus is the same as hunger in a mammal. That the same psychological state could be present in such diverse lineages would be quite surprising given mind-brain identities, but not at all given the possibility of multiple realization. However, we contend, this argument runs the risk of conflating behaviors with their psychological causes. In particular, it appears to conflate the state that produces food-acquiring behavior (viz., a psychological cause) with the food-acquiring behavior itself.

---

[6] A large shark will die when dragged ashore because with no rib cage to protect its internal organs, it will be crushed under its own weight.

Suppose we grant that both octopuses and, say, otters exhibit a similar behavior, i.e. both engage in foraging behavior.[7] However, we doubt that the psychological causes of this behavior in each kind of organism are similar. This is important because the multiple realization thesis concerns psychological processes rather than behaviors, and psychological processes are often (but not always) causes of adaptive behaviors. Thus, even if the foraging behaviors of octopuses and otters really are the "same" this provides no guarantee that the psychological causes of these behaviors are the same. Perhaps the internal causes of foraging behavior in an octopus are not psychological at all.[8]

Of course, calling into question whether the cause of octopus foraging behavior is psychological might, from a certain perspective, seem to beg the question against early functionalists such as Putnam. On the view Putnam defended, a psychological state, such as hunger, was characterized as that state which, in response to deprivation of food, causes foraging behavior (along with other psychological states). Yet, in retrospect, this conception of a psychological process can hardly do justice to realistic explanations of foraging behaviors in different species. Philosophers and psychologists rightly reject the idea that the state which leads a flatworm to forage must be the same, or even similar, to the state that causes foraging in parrots, koalas, and chimpanzees. Even single-cell organisms engage in "foraging" behaviors. No doubt, in most kinds of organisms some internal process is the cause of foraging behavior; but it is probable that some of these processes might be psychological and others not. That foraging is caused by an internal psychological state of hunger is certainly not a universal truth about organisms.

To reiterate: When we ask about multiple realization we are asking about the relationship between brain processes and psychological processes. These psychological processes are the causes of behaviors. It is uncontroversial that any behavior might have many causes. Even

---

[7] In truth, even this claim makes us uneasy. Clearly, the kinds of movements by which an octopus gathers food is nothing like those by which an otter gathers food. Only at the coarsest of descriptive grains do the two organisms engage in the same behavior (cf. Bechtel and Mundale 1999). We don't deny that octopuses and otters both forage, but we are reminded that how a set of movements is characterized often already embodies assumptions about its etiology and purpose, and sometimes about whether it is guided by cognitive processes or not.

[8] Remember, advocates of multiple realization are cognitivists, not behaviorists. There is more to hunger than simply being caused by lack of food and causing foraging.

complicated behaviors that have psychological causes in some creatures could have causes that are not psychological in other creatures. Even if convergence makes it probable that all creatures will engage in certain very general kinds of behavior, such as the acquisition of nutrients and the avoidance of injury, there is no reason to suppose that those behaviors would invariably be caused by psychological processes in all creatures.

What distinguishes psychological causes from non-psychological ones? Our present purpose, which is to cast doubt on the inference from convergence in behavior to convergence in psychological state, does not require that we answer this question. Whatever the nature of a psychological state, the point stands that behaviors can be similar without any similarity in psychological causes, just as the shapes of the shark and the dolphin can be similar despite differences in their construction. Perhaps some story involving informational content might serve to distinguish psychological from non-psychological states. Or perhaps psychological states are accompanied by some kind of qualitative feel. Whatever the answer, varieties of functionalism that would automatically attribute the very same psychological states to octopuses, otters, and mosquitoes strike us as far too permissive. If convergence is to provide any kind of evidence for the multiple realization of psychological states, it had better get the target right. It must be convergence in psychological processes and not just the behaviors they cause. Advocates of multiple realization need to assure us that the convergence that so impresses them goes beyond mere behavioral convergence and includes convergence on psychological causes of common behavior.

But, even if we could be assured that humans, octopuses, and otters have common psychological states, would the observation of such convergence be evidence in favor of multiple realization? Returning to our likelihood framework will help to illuminate the problem we foresee. Convergence says that we observe the same psychological states across members of numerous species. The argument that such an observation favors multiple realization over the identity theory goes like this:

(C1)   Given Multiple Realizability, we expect to observe convergent evolution of psychological processes and mechanisms.

(C2)   Given Identity Theory, it would be surprising to observe convergent evolution of psychological processes and mechanisms.

(C3)   Thus, Multiple Realizability has a higher likelihood than Identity.

The suggestion is that evolutionary considerations would make convergence more surprising if the Identity Theory is correct than if Multiple Realizability is correct. The idea, we suppose, is that natural selection cares more about the behaviors an organism exhibits than the "behind the scenes" causes of the behaviors.[9] Therefore we should expect to observe different causes of those behaviors in different creatures, and those would constitute different realizers of the same psychological states or processes.

Granting for the moment that these different causes of behavior count as psychologically the same, we wonder: Why would Multiple Realizability make convergence of this sort less surprising than would Identity Theory? As Putnam himself observed, convergence produced the same kind of realization of eyes in cephalopods and mammals. Why not expect convergence to result in the same kinds of realizers across species with psychological traits? What is it about psychological traits, in contrast to eyes, that makes them unlikely to be realized in the same way as a result of convergent evolution? If natural selection has contrived to produce similar psychological states in various creatures, and not merely similar behaviors, then perhaps many of those psychological states are similarly realized. The argument that persuades us pays careful attention to evolutionary *constraints* (Shapiro 2004). A constraint, in the present context, is a limitation on the number of ways a given function might be realized. We shall have more to say about the source of these limitations shortly. For now, a few examples will help to elucidate the idea of a constraint.[10]

Suppose you wish to build a craft that will carry you across a lake. Archimedes' Law, which tells us that the upward force on an object in a fluid is equal to the weight of the fluid that it displaces, constrains the kinds of realizers of such a craft. This constraint requires that the craft you build to weigh less than the amount of water it displaces. That is, its average density must be less than that of water. Any realizer that fails to meet this constraint will sink. Watercrafts realized in wood can satisfy this constraint, as can those made from metal if designed in such a way that their average density meets the constraint that Archimedes' Law imposes.

---

[9] Obviously this is an oversimplification. Some causes will carry higher evolutionary costs than others.

[10] For further discussion of convergence and constraints, see Couch (2005), Wimsatt (2007), and Towl (2012).

As an example of a different sort of constraint, suppose that you are a clockmaker and have fashioned a clockwork with gears of various sizes. The mechanism is nearly complete. You need to insert just one more gear. The existing collection of gears imposes constraints on the realizer of the final gear. The teeth of the final gear must be appropriately sized, so that they fit between the teeth of the surrounding gears (and those of the surrounding gears fit between them). Failing this constraint, the final gear would not realize the function that it was intended to play.

From these examples, we wish to draw several lessons. First, constraints on realizers might take at least two forms. We shall refer to *nomological constraints* as those that follow from laws of nature. Of nomological necessity, a watercraft must be less dense than the fluid on which it is intended to float. In contrast, the realization of the gear in the clockwork is subject to constraints not (only) from natural law, but from those that result from the special circumstances that the other components of the clockwork create. Because the surrounding gears have the number of teeth that they do, and because these teeth are a certain distance apart from each other and have a certain length, the final gear, if it is to serve its function in the clockwork, must be realized like *so*. *Circumstantial constraints*, unlike nomological constraints, are dictated by historical happenstance. Given that the construction of the clockwork took the course that it did, the final gear had, by circumstantial necessity, to have *n* number of teeth, with dimensions *x*, *y*, and *z*.

Although we have distinguished nomological from circumstantial constraints, the distinction is not always clear-cut. For instance, organisms with smaller bodies are more subject to dehydration than organisms with larger bodies. This suggests a constraint: Smaller organisms must have strategies to ward off dehydration. But does this constraint reflect particular circumstances or natural law—or, even, *mathematical* constraints? Given the circumstance of a small body, the risk of dehydration imposes a constraint. But, of course, it's because of the lawful relation between surface area and volume that small organisms have more "outsides" relative to their "insides," and that is what renders them vulnerable to dehydration (Vogel 2013: 45–7). So, it looks as though whether we choose to characterize the constraint on small-bodied organisms as circumstantial or nomological depends on how we describe the situation. Given the circumstance of having a small body, an organism must find ways to combat dehydration. Or, given the lawful relationship between surface area and volume—that surface area "squares" while volume "cubes"—small-bodied organisms must have strategies to avoid dehydration.

Importantly, circumstantial and nomological constraints often accumulate as a result of each other's presence (cf. Shapiro 2004, Wimsatt 2006). Perhaps, for instance, an organism relies on endothermy to regulate its body temperature. This circumstance is contingent; temperature regulation might have been accomplished in some other way, as it is in cold-blooded organisms. But, given that the organism is endothermic, it must, by nomological necessity, acquire large amounts of energy. (Endothermic organisms need to consume roughly ten times as much food per unit weight than cold-blooded organisms.) But, given circumstantial facts, such as that the organism evolved in an environment where food had to be hunted, it might be constrained to spend large amounts of time searching for food. But locomotion, by nomological necessity, causes the organism to heat up, thus placing new burdens on its temperature-regulation system, the solution to which will be constrained by other features that the organism happened to possess. What we see, then, is an interplay between circumstantial and nomological constraints where one leads to another, which in turn, leads to the other, and so on.

Thinking about this cascading series of constraints brings us to the next point. We assume that, as a practical matter, when systems grow in complexity they frequently face increasing numbers of constraints. Admittedly, this principle is imprecise. We shall not attempt to characterize complexity except to suggest that one aspect of a system's complexity involves its number of parts and variety of kinds and arrangement of parts. Motivation for the principle rests on the plausible idea that as the internal complexity of a system increases, the demands placed upon any new addition to the system usually become more exacting. The more complicated a clockwork—the more gears, springs, oscillators, and so on that it contains—the more constrained additions to that clockwork will be.[11] This is because any new component to a clockwork must fit with the existing components:

[11] Herbert Simon (1962, 1996) uses an example involving two watchmakers who produce equally complex but differently assembled watches to argue that the conditions on evolution, like those on efficient design, favor the production of "nearly decomposable" systems ("hierarchical" or, approximately, modular). He goes on to argue that in such systems "the efficiency of one component (hence its contribution to the organism's fitness) does not depend on the detailed structure of other components" (Simon 1996: 193). Some commentators have suggested to us that this implies that for evolved systems, such as known cognitive

Properties are not free to vary independently. If material objects or configurations are responsible for the upper-level phenomena, just a few observed correspondences provide a strong filter for arrangements or configurations of objects and interactions that could produce these results. These articulated constraints are what make mechanisms such powerful explanatory tools in the compositional sciences. (Wimsatt 2006: 454)

The more components present, the fewer degrees of freedom available to a new component. If the working parts of brains are highly multifunctional, e.g., if there is a great deal of neural reuse, then the complexity of brains isn't much of an obstacle to their evolving new capacities. But we expect convergent evolution often to find similar realizers for similar capacities, for it is repeatedly drawing from the same stock of components, which, in turn, must respond to the same or similar constraints (see Figure 3.1 for graphic illustration of this idea).

We assume that the systems that realize psychological capacities are very complex, involving mechanisms with numerous and varied components. When combined with the assumption that complexity produces constraints, we can see why complexity casts doubt on Block and Fodor's confidence that convergent evolution will lead to the multiple realization of psychological capacities. For if brains are complex, their components are constrained; therefore similar psychological capacities will often have similar realizers. For instance, if two distinct kinds of organisms have very similar perceptual abilities, where these abilities evolved independently, we should not be surprised to see these abilities realized in a very similar way.[12] Thus, the complexity of psychological capacities favors the Identity Theory hypothesis over Multiple Realizability.

systems, there are relatively few constraints on their neural realizers because there are relatively few interactions between subsystems compared to the interactions within a subsystem. We regard as an open question the extent to which the brain is modular in a way that favors Simon's suggestion. Insofar as Anderson (2010, 2014) is correct, many neuron and neuronal circuits participate in multiple cognitive processes, calling into question modularity. And, even if the brain displays modularity in peripheral systems, as Fodor (1983) argues, it might not in the "central" systems where higher thought takes place. Finally, we take our example of clockworks to suggest that the peculiarities of components in a system, regardless of whether they have clearly defined interfaces with each other, may constrain the kinds of additional components that can be integrated into the system.

[12] The stipulation that the capacities evolved independently is crucial, because if they simply descended from a common ancestor then their similarity can be explained without appeal to the complexity of the system.

In general, we contend, the presence of constraints together with the complexity of psychological capacities suggests that these capacities will often have relatively few kinds of realizations, and sometimes just one kind. Nevertheless, because creatures' capacities differ and the severity of the constraints working on the realizations of these capacities no doubt differ as well, we must remain open to the idea that some—even many—psychological traits may be multiply realizable. This is an acknowledgment of some importance, because it brings into clearer relief the extent to which multiple realization, despite lip service to the contrary, has often been treated as an a priori truism. Proponents of multiple realization have often adopted the attitude that *all* psychological states or capacities can be multiply realized. However, no clear basis for this confidence has, to our knowledge, ever been articulated. Why should the fact that the kind *pain* is multiply realized (if indeed it were) predict anything about whether the kind *visual perception* is multiply realized, or whether the belief that philosophers drink too many martinis is multiply realized? Does the fact that corkscrews can be multiply realized entail that hearts can be as well? We do not see why. Daniel Weiskopf (2011), for example, argues that convergence can produce multiple levels of what he calls "functional unity" for evolved special sciences kinds (see also Wimsatt 2013). But that is fully compatible with some convergently evolved psychological kinds being identifiable with neural kinds.

Do psychological kinds cross-classify neuroscientific kinds? Evolutionary considerations such as convergence do not predict a univocal answer, contrary to Block and Fodor's suggestion. Evolution is an engine of contingency, not the engineer of interchangeable parts.

At this point it is worth taking stock of our commitments, and of the challenge that we have set out for the advocates of multiple realization. The identity theory holds, according to us, that there are important and explanatory mind-brain identifications. This is a significantly more modest view than has been frequently attributed to identity theorists, but still one at odds with realizationist or functionalist theories according to which multiple realization is so ubiquitous that no interesting or

We are indebted to Rosa Cao for discussion of constraints and multiple realization. Cao also raises the possibility that capacities that appear to have independently evolved may turn out to be due to the (re)activation of a common ancestral capacity or developmental module. Something like this could well be the fact about the electroreceptive capabilities of the South American and African electric fish, for example, which evolved from non-electroreceptive species but have common electroreceptive ancestors (Kawasaki 2009: 588).

useful mind-brain identifications could be discovered. We do not deny that multiple realization occurs; we deny only that it occurs to an extent that renders the identification of psychological kinds with brain kinds impossible or unimportant. And we demonstrated in earlier chapters that some *prima facie* examples of multiple realization turn out to be cases in which explanatorily valuable mind-brain identities can be found. We are enthusiastic fans of both plasticity and convergent evolution. But we doubt that they provide reasons to estimate that universal Multiple Realizability is a better hypothesis than our modest Identity Theory.

The present chapter and the next are concerned with the idea that the evidence we have offered for useful mind-brain identities is misleading. The worry is that we are letting ourselves be hoodwinked by contingent features of actual terrestrial creatures and not features of psychological and neural systems considered in general, i.e., setting aside the peculiar contingencies of how they have actually evolved on Earth. The idea is that there are some universal principles concerning brains (e.g., their multipotentiality, plasticity, or degeneracy) or organisms (e.g., that their capacities can be produced by convergent evolution) that make it overwhelmingly likely that every cognitive process or capacity is in principle multiply realizable regardless of the sorts of accidental similarities we observe among cognitive systems around here. But keep in mind that, given that we acknowledge the evidence of some multiple realization, insisting only on the existence of *some* important and explanatory mind-brain identities, our opponent must show that *every* cognitive process or state is multiply realizable so that no useful kind identities can be found.[13]

Up to this point we have focused on whether plasticity, degeneracy, or convergence give us reasons to think that the different processes that produce certain behaviors are, upon closer examination, relevantly different. But that's only one part of the Official Recipe. For the fact that degenerate or convergently evolved neural systems can produce behaviors similar to those produced by canonically psychological systems does not demonstrate the multiple realization of psychological capacities. As far as degeneracy and convergent evolution are concerned, it could equally well be the case that those behaviors might not be produced by *cognitive* mechanisms at all. Degeneracy and convergent evolution

---

[13] It's reasonable to wonder why our opponent should be so committed to such a strong view.

might be reasons for expecting multiple neural causes for some effects. But, short of embracing behaviorism, one can't assume that common stimulus-response pairings reveal the presence of common psychological processes. This hidden assumption is particularly striking in the convergence argument. Natural selection can stumble into various solutions to ecological challenges, and various ways to cause similar behaviors. But why think that all of those solutions will involve psychological causes? And, even granting that the proximate causes of a certain behavior include one or more psychological processes, the burden falls on the proponent of multiple realization to show that the various realizers of this process are in fact realizers of the *same* process—rather than realizers of distinct psychological processes that produce the same behavior. Degeneracy and convergence alone don't provide evidence of this sort.

What the defender of multiple realization needs to provide is not more information about brains or evolution, but some general principle about the nature of psychological processes that guarantees, or at least suggests, their multiple realizability. And this is where the analogy between minds and computing machines comes into play.

5 Machine Minds

The last consideration that Block and Fodor raise is that multiple realization is to be favored over identity because of the "conceptual possibility" of machines psychologically similar to organisms despite lacking organism-like brains. They claim:

it seems likely that given any psychophysical correlation which holds for an organism, it is possible to build a machine which is similar to the organism psychologically, but physiologically sufficiently different from the organism that the psychophysical correlation does not hold for the machine.

(Block and Fodor 1972: 161)

The argument these considerations suggest fits easily into our likelihood framework:

(AI1) Given Multiple Realizability, we expect to observe artificially intelligent machines that are psychologically similar to human beings.

(AI2)   Given Identity Theory, it would be surprising to observe artificially intelligent machines that are psychologically similar to human beings.

(AI3)   Thus, Multiple Realizability has a higher likelihood than Identity.

As it stands, we do not find this argument at all compelling. However, it does invite reflection on a more persuasive reason to accept the possibility of multiple realization. We'll turn to that reason in the following chapter. For now, we'd like to offer a brief reply to the argument above.

First, the argument can hardly be convincing without actual observations of artificially intelligent beings that exhibit a human-like psychology. It does no good to claim that one hypothesis should be favored over another because it predicts an observation that the other does not when the observation has in fact not been made! The artificial intelligences that Block and Fodor claim to be a possibility remain only science fiction, nearly fifty years after the time of their writing. Matters might have gone otherwise, we suppose. In the time since Block and Fodor made their argument, an artificial being with human-like psychological capacities might have been invented. Had this occurred, we would then be in a position to investigate its hardware and determine whether, by our Official Recipe, it provides an example of multiply realized psychological capacities. But history has not gone this way. No artificial intelligence in existence today displays anything like normal human psychological capacities, and so we're left with only the promise of future artificial intelligence. And, of course, that is not enough to make us favor Multiple Realizability over Identity Theory. The possibility of evidence is not evidence of a possibility.

Second, given that we are not being directed to observe any actual examples of artificial intelligence, the stipulation that these intelligent non-human systems are "artificial" does no work. The first premise might have instead been, "Given Multiple Realizability, we expect to observe natural selection creating other organisms that are psychologically similar to human beings"—which was the convergence argument. Or, "Given Multiple Realizability, we expect to observe intelligent extra-terrestrials that are psychologically similar to human beings"—which again points to possible but non-actual observations. When Block and Fodor appeal to the possibility of artificial intelligence as an example of something that *could* be psychologically similar to a human being while also differing in physically relevant ways, they might have mentioned instead pretty much anything. Adverting to "a machine which is similar to the organism psychologically" is just a way of referring to something or other that is psychologically like human beings while being physiologically different.

Once we see how easy it is to recast Block and Fodor's argument from the possibility of artificial intelligence in terms of the possibility of anything at all that might be psychologically similar to human beings (while also physiologically distinct from us), it becomes apparent that the argument is question-begging. Why accept, as Block and Fodor seem to do, that the psychological beings they imagine us building will differ in relevant physical ways from human beings? We agree that artificial and alien cognizers seem possible, at least in the sense that it is an open question whether we will ever build or discover any. Yet that possibility only counts in favor of multiple realization if we assume that those cognizers will differ relevantly from human beings in how their cognition is realized. However, we do not share Block and Fodor's confidence that we can know in advance that the processing units of human-like artificial intelligences—or the brains of extra-terrestrials, or future products of natural selection—will differ relevantly from the brains of human beings. What evidence do we have for thinking so? That, after all, is the question we've been asking: Are there general reasons to expect that cognitive processes are all multiply realizable—possibly multiply realized? Block and Fodor seem to be adopting as a background assumption that it just stands to reason that human-like artificial or alien intelligences would be realized in nothing like a human brain. But, of course, this stands to reason only if you already accept Multiple Realizability. So this possibility cannot be a non-question-begging reason to believe that Multiple Realizability is more likely than the Identity Theory.

For the reasons just given we find little to recommend Block and Fodor's final argument for Multiple Realizability, taken at face value. However, as we mentioned, the argument does suggest another reason why some might be attracted to the idea of multiple realization. The development of artificial intelligence as a research area occurred alongside the elaboration of the *computational theory of mind*. Even if one agrees with us that appeals to possible artificial intelligences cannot provide evidence for Multiple Realizability, one might think of the attractiveness of this possibility as rooted in the belief that minds are essentially computers—that mental processes are computational processes. This would be the start of a very different kind of argument for Multiple Realizability than the one we just discussed, but, we suspect, it is an argument that has convinced many to accept the possibility of multiple realization.

The *computational theory of mind* approaches psychological explanation from the perspective that mental systems are computational systems or—perhaps more generally—information-processing systems.[14] The idea, roughly, is that the psychological laws (or taxonomic features, more generally) that are individuative of psychological states and processes will be purely computational or information-processing laws. According to this view, the taxonomy of psychological processes is not constrained at all by behavioral or neuroscientific details of their realization in terrestrial creatures, but will instead be characterized entirely in terms of the mathematical or syntactic features of computation or information processing. Perhaps adherence to the computational theory of mind explains why Block and Fodor are confident that both behaviorism and the identity theory are "insufficiently abstract."

The issues surrounding the computational theory of mind and its relationship to the multiple realization hypothesis are sufficiently complicated to warrant a chapter unto themselves. We thus leave behind our discussions of plasticity, convergence, and artificial intelligence and turn in the next chapter toward a detailed examination of whether cognitive processes are computational in a way that implies their multiple realizability.

---

[14] Piccinini (2015) argues that it is a mistake to conflate computation and information processing, and that the two contribute to cognitive explanations in different ways; but he also acknowledges that, as a matter of fact, they frequently run together in the cognitive sciences and philosophy. We will continue to use the two as a pair, and Piccinini acknowledges that computational explanations of cognition typically assume that cognition involves processing of semantic information. He also argues that there is a generic notion of computation such that every information-processing system is a computer. What is important for our purposes is that, in Piccinini's framework, the features relevant to the question of multiple realization belong essentially to computers and only accidentally to information processors. See also Piccinini and Scarantino (2011), Piccinini and Bahar (2013), and Piccinini and Shagrir (2014).

8

The Computationalist Argument for Multiple Realizability

1 Computational Explanation and Task Analysis

The fruitfulness of computational and information-processing theories in the cognitive sciences has certainly impressed many realization theorists. Paradigms of the computational approach include Noam Chomsky's work on generative grammar (1957), Newell and Simon's work on problem solving (1972), David Marr's computational theory of vision (1982), and Seidenberg and McClellands' work on word naming (1989, 1990).[1] Jerry Fodor famously claims that "the only psychological models of cognitive processes that seem even remotely plausible represent such processes as computational" (1975: 27). This is sometimes called the "only game in town" argument for the computational theory of mind, because Fodor takes his cue from Lyndon B. Johnson's pronouncement, "I'm the only President you've got" (Fodor 1975: 27). Moreover, Fodor is adamant that computational descriptions of mental processes should not be interpreted as mere *façons de parler*—we should

---

[1] Some influential surveys include Pylyshyn (1984), Stillings et al. (1987), and Von Eckardt (1993).

take literally the hypothesis that minds are computational systems. Fodor and many others have supposed that minds must be multiply realizable because they are computational.

The argument for the multiple realizability of the mental that we are now considering goes something like this: The best explanations in the cognitive and psychological sciences are those that treat cognitive and psychological processes as computational. That is, the *computational theory of mind* is true. If the computational theory of mind is true, psychological states and processes are individuated by "abstract" laws relating computational and informational states and processes, entirely blind to facts about their physical realization. For this reason we should be confident that the taxonomy of psychological kinds cannot be brought into alignment with the taxonomy of neuroscientific kinds. Thus we can be sure that psychological processes are *multiply realizable* even if we do not observe frequent *multiple realization* among terrestrial creatures. What we observe, rather, is an indicator of multiple realizability—namely, widespread use of computational explanation in the mind sciences.

It seems to us that a great many philosophers think that mental states must be multiply realizable for reasons like these. They accept something like the computational theory of mind and they take for granted that the computational theory of mind entails that mental states are multiply realizable. Computational processes don't have to be implemented in brains, after all. But we believe that the line of reasoning rehearsed above is mistaken.

Consider what is necessary for the computational theory of mind to rule out significant and explanatory mind-brain kind identities: First, it must be true that psychology and cognitive science offer computational explanations, in the sense of explaining cognitive and psychological behaviors by appeal to internal computational processes. Second, these computational explanations have to be ontologically committing—we must think that they take psychological states and processes to be identical to abstract and multiply realizable processes, not merely to be describable as such. Put another way, the abstractness of the explanations has to be a feature of the psychological states, not merely a feature of our representations of them. Third, these real computational processes have to be sufficiently "abstract" that they are multiply realizable in principle. That is, the computational kinds must be guaranteed to cross-classify neural kinds in the ways described by our Official Recipe.

And, finally, these abstract and internal-computation-invoking explanations have to be sufficiently prevalent that we can rule out in principle any explanatorily important or useful kind identities between brain and psychological kinds. We don't deny that some explanations in the cognitive sciences fit the bill. But many do not, and that leaves room for our modest identity theory.

Begin with the purported observation that the cognitive sciences explain intelligent or psychological behaviors by appeal to internal computational processes. It would be foolish to deny that cognitive and psychological journals and textbooks are full of apparently computational explanations. But there is a great deal of confusion among philosophers and cognitive scientists alike about what makes an explanation computational. The groundwork for this confusion was laid by Marr himself, with his famous distinction between three levels of analysis of cognitive systems: Computational, algorithmic, and implementational. According to Marr, the "computational" explanation is a task analysis, it tells us what the system does and why. The algorithmic explanation tells us what processes the system uses to produce its behaviors, and what representations are manipulated in doing so. (Marr assumed that cognitive algorithms are operations on representations.) And the implementation explanation tells us about the physical properties of the systems that "realize" the algorithmic procedure, be they silicone, neural, or whatnot.

The important observation for our present purposes is that there is a familiar kind of "computational" explanation—i.e., Marr's style of task analysis—that is actually silent on the nature of the processing that occurs within the system; it merely tells us what behaviors or representations the system produces and why. Yet as we emphasized earlier, multiple realizability involves the cognitivist commitment to the reality of psychological processes that are shared across multiple systems or kinds of systems, and that produce the distinctively cognitive behaviors. This is why we require more for multiple realization than multiple causes of similar effects. From the fact that a system has a Marr-style "computational" task analysis, nothing follows about whether the processes that carry out the task are themselves multiply realized or realizable. We illustrated this interpretive challenge in Chapter 3 with the example of honeybee face recognition. There we saw that the superficially computational explanation of the honeybees' ability to recognize faces—"configural processing"—turned out to be merely a way of describing their capacities: They are sensitive to configurations of features, not merely to isolated face-like features (Avarguès-Weber et al. 2010). In this case, it is a live question for the researchers whether the processes that produce the behavior should be explained as either cognitive or computational (Dyer 2012). So an apparently computational claim, viz., that bees recognize faces using configural processing, in fact merely describes the relationship between stimuli and behavior. Configural processing explains the bees' behavior in that it provides a task analysis without staking claims about whether the intervening processes are algorithmic, much less computational in a more robust sense, e.g., representational.

Furthermore, many examples of "computational" explanation in the cognitive sciences are like this. Take Marr's canonical computational theory of early vision, according to which cells in the retina identify luminance boundaries in the retinal image by an algorithmic transformation on the retinal representation (viz., Laplacian of the Gaussian, $\nabla^2 G$), and then detecting zero-crossings in the transformed image (Marr 1982). But as Marr himself admits: "From a physiological point of view, zero-crossing segments are easy to detect without relying on the detection of zero values, which would be a physiologically implausible idea" (1982: 64). He goes on to explain that a different algorithm is more "physiologically plausible" but the important point is that sometimes, even at the algorithmic level, Marr seems to think of the system as implementing the "algorithm" just in case it has the right input-output mapping. That is, Marr seems to accept that sometimes a sufficient condition for positing a given algorithm is that the system successfully "computes" a given function. The instruction to "transform retinal input by $\nabla^2 G$ and locate zero crossings" does not describe components and processes in the retinal system, but this does not prevent Marr from giving that as the algorithmic explanation of this stage of visual edge detection. Thus it seems that the only sense in which the retina implements the specific algorithm described by Marr is that retinal cells "compute" the task that the algorithm is supposed to explain (Polger 2004b; see also Orlandi 2014). But, in this context, "compute" means little more than "produce the specified output."

Even when the computational task description involves taking representational inputs or producing representational outputs, attribution of an intervening process of representation manipulation is not a foregone conclusion. Kosslyn and Hatfield (1984) describe a spring balance that can be used to compute the sum of two weights, simply by resting the weights on its pan. A computational analysis of the device would match inputs to outputs, revealing the device to be an adder. Although the device can correctly be described as computing the sum of the two weights, producing a representation of the sum as output (by pointing with its arrow to a numeral), the compression of the spring by which it does so is not an algorithmic process in Marr's sense, for it does not manipulate representations. Moreover, even if one were willing to contort the notion of "algorithmic" in order to make it apply, the idea that such a description is ontologically committing remains an even further leap. We contend that the spring scale illustrates a system that can be described as computing a function—addition—resulting in a representation—the sum—but does so in a non-algorithmic fashion, i.e., one that does not involve the manipulation of representations.

We draw two lessons from these examples. Most importantly, some apparently computational explanations in the cognitive sciences are, on closer inspection, best understood as precise descriptions of the task or behavior to be explained, e.g., honeybees exhibit "configural processing." This is consistent with these explanations or descriptions being "computational" in Marr's sense, but does not imply anything at all about the multiple realizability of the processes that carry out the task, i.e., that produce the behavior. And this is true even when the computational task descriptions posit representations as inputs or outputs. The processes that intervene between the inputs and outputs may involve algorithms, but they may be better characterized entirely in physiological or (as in the case of the spring balance) mechanical terms.

Second, even in those cases where cognitivist explanations specify internal algorithmic processes, sometimes those algorithms are presumed to be implemented whenever their total inputs and outputs match those of an algorithm. This remains true even when no physical structures or events correspond to the processing components or steps specified by the algorithm, as in the case of retinal cells that Marr says detect values in the $\nabla^2 G$ filtered retinal image. In such examples as these, the apparently algorithmic explanation may provide a way of modeling the task that the system does, but should not be taken to describe the actual intervening process. In Michael Weisberg's terms, they have *dynamical fidelity* but not *representational fidelity*; they do not characterize the actual components and operations of the system (Weisberg 2013: 41; see also Miłkowski 2015).

Computational explanations with these features provide no reason to deny that the taxonomy of cognitive processes can be aligned with the taxonomy of neuroscientific processes. They either tell us nothing at all about internal processes, or they are ontologically silent about the nature of those processes. In short, before appealing to the computational theory of mind in order to defend multiple realizability, one must justify a very particular conception of the commitments of such a theory, commitments that cognitive psychology hardly makes *de riguer*.

2 Abstractness and Multiple Realizability

Now we want to set aside the concern that computational models don't model the real processes that produce the inputs and outputs, and instead focus on the idea that a process counts as genuinely *computational* just in case it has an appropriate abstract or computational model. Is there an argument for the in-principle multiple realizability of cognitive processes on the grounds that such processes have computational models or descriptions?

Not surprisingly, we favor a negative answer to this question. The first problem concerns the "abstract" nature of computational descriptions and explanations. Many philosophers take this abstractness to suffice for multiple realizability but, we shall argue, it does not. Secondly, once we get a better handle on the right kind of abstractness to motivate multiple realizability, we might wonder about the prevalence of computational explanations in psychology. Staunch defenders of computationalism cast it as "the only game in town," but this claim sells short a host of other approaches to understanding cognition. Even if some cognitive processes are correctly described as computational, we doubt that all must be; and so we find little reason to expect the universality of multiple realizability.

Consider the much touted abstractness of cognitive and psychological explanations, which ensures that they can apply to systems of a variety of compositions and organizations—"copper, soul, or cheese" as we quoted from Putnam earlier (1973 in 1975: 292). This is an important feature of the nature of cognitive systems, according to the computational theory of mind. Yet it is not generally true that the abstractness of scientific models corresponds to any special ontological features of their objects. Abstract models leave out some information about particular systems, and because they leave out information they can be applied to systems that differ with respect to those omitted details. It would be a mistake to infer from the fact that psychologists make use of abstract models that the phenomena so modeled permit multiple realizations. For instance, from the fact that the relatively "abstract" or "topic neutral" (Smart 1959) characterization of H_2O as *water*, or as *the cool, clear, potable, liquid that rains from the sky*, nothing at all follows about the multiple realizability of that kind of stuff, H_2O. Scientific models often apply to many different systems because they omit information about the distinguishing features of those systems, not because they positively assert that the things they model are abstract in some special way. In fact their abstractness is sometimes a signal that they are not ontologically committed to the entities that they model, at all.

Remember that the view we're considering at the moment is that the "abstractness" of computational descriptions and explanations is just a matter of the fact that these descriptions can apply to systems of various physical compositions and structures. On this view, the claim that some physical process is computational is merely the claim that it is in some ways similar or isomorphic to a computing process—that they are described by the same models. One might model a river system as a system of pipes, or as a geometric structure, or as a series of decisions. "Water seeks the path of least resistance," every hiker knows. Such models can be quite useful in understanding and predicting the behavior of a system; perhaps even in explaining some aspects of the behavior of a system. But a river system is not in fact a system of pipes or decisions. River systems can be modeled as computational or information-processing systems, as well. That the flow of water through a river system can be modeled as a computing process does not show that river systems are in fact computing machines, or that they have all (and only!) the properties mentioned in their computational descriptions, i.e., models. The utility of abstract models, including computational models, rests on their ability to facilitate both experiment and understanding, but it is generally a mistake to take the abstractness or generality of the models to represent a fact about the phenomena being modeled, viz., that the phenomena are themselves in some way abstract. Generally, that a model omits some facts about the nature of the realizers of a process reveals little about the relationship between the phenomenon and its would-be realizers.[2] Why expect anything different in the specific case of computational models of cognition?

Advocates of realization theories, particularly in their computational versions, have tended to think that the abstractness of computational descriptions of cognition carries more ontological significance than the abstractness of other scientific models. This idea seems to be related to the idea that a system is computational simply in virtue of having a computational description.[3] If being a computational system is just a matter of having a certain kind of description, and that description can apply to systems of various physical compositions and structures, then computational systems can have various compositions and structures. In such a case, the presence of computational models in the cognitive sciences guarantees multiple realizability.

According to this line of reasoning, the abstractness of computational models of cognition is not the run of the mill abstractness of scientific models. Computational explanation of psychological processes differs importantly from, e.g., the computational modeling of river systems.[4] The reason is that being a computer, unlike being a river, really is just a matter of having a special kind of description. And that's why the computational theory of mind implies that minds are multiply realizable.

If this is the line of reasoning from computationalism to multiple realizability, computationalists must tell us what it is about cognition, in contrast to rivers, that forces us to accept that computational explanations are ontologically committing. And there is a familiar answer to this challenge. Fodor (1975, 1985), as well as computationalists like

---

[2] Suppose that someone thinks the susceptibility of a phenomenon to modeling using a certain kind of abstraction is indicative of the deep nature of the phenomenon, after all? Then we can point out that the omitted features are apparently irrelevant from the point of view of explaining the phenomenon being modeled. So the fact that these irrelevant features can vary without changing the phenomenon would not be even prima facie evidence of multiple realizability according to our Official Recipe.

[3] This theory of computation is widely rejected, but let us ignore that difficulty for the moment.

[4] Gualtiero Piccinini (2007, 2015) distinguishes those things that have computational models, e.g., rivers and digestion, from those that have computational explanations, viz., computers. John Searle seems to have had something like this point in mind when he insists on distinguishing simulation from replication or emulation (1980). In contrast to Piccinini's regimentation, we use the expressions "computational explanation" and "computational model" interchangeably, so that the ontological commitments of computational explanations can be open to question.

Pylyshyn (1984) and Haugeland (1981, 1985), argued that psychological processes are computational in the specific sense that they involve manipulations of representations. The representationalist version of the computational theory of mind contends that the isomorphism of cognitive and computational processes reveals something deep about the nature of cognitive processes rather than merely providing a tool for studying them. According to this version of the computational theory of mind, computing is a matter of manipulating representations and the nature of representations is such that whatever can be modeled as a representation is *ipso facto* a representation, so being a computing system really is just a matter of having a special kind of description, after all. This accounts for the special abstractness of computational models of cognition and explains as well their multiple realizability.

If the representation or symbol manipulation version of the computational theory of mind is correct, then we suppose it would imply that psychological states are multiply realizable in whatever ways that symbol manipulating processes are. This is the real crux of Fodor's (1975) "only game in town" argument—that the only credible explanation of cognition treats it as a process of manipulating symbolic representations.

But if the imperative for representationalist computationalism is the reason to regard psychological states as multiply realizable, then it is a poor reason. This representationalist version of the computational theory of mind is plainly not the only game in town. It's simply not true that representational and symbol manipulation explanations are the only "intelligible" or "plausible" contenders in the cognitive sciences. Numerous critiques of the "only game in town" argument have been given over the years. The first round of objections, widely discussed in the 1980s and 1990s, concerns whether connectionist networks offer a genuinely alternative model of cognition. Connectionist models reject the "symbol manipulating system" conception of both computing and cognition in favor of systems involving numerous, simple, and non-symbolic processing "nodes" or "neurons" that do not follow a stored program (for discussion of such models, see McClelland et al. 1986, Rumelhart et al. 1986, Smolensky 1988, Bechtel and Abrahamsen 1990, Hatfield 1991, Clark 1993). Such networks are surely computing machines in one sense; and they can be used to model or explain a wide range of human and animal behaviors.

Nevertheless, although connectionist networks do not involve the manipulation of symbols, or, perhaps less controversially, do not obviously contain symbols that exhibit the compositionality characteristic of "good old-fashioned" computation (Haugeland 1985), they remain computational in the sense that they process information. (More on information processing below.) And, significantly, they are "abstract" in the sense that they can be built out of most anything. Indeed, connectionist networks are almost always simulated or emulated using traditional symbolic computing machines—thereby illustrating the fact that their operation is suitably independent of the details of their realization. So while connectionist networks may be an alternative to the kinds of computational systems that Fodor imagined, they are not an alternative in a way that undercuts the part of his argument that concerns us. If cognitive processes took place in abstract connectionist networks, that would be a reason to accept their multiply realizability. So if the choice were only between symbol manipulating and connectionist systems, some sort of computationalism would still be the only game in town.

But those are not the only alternatives. We will highlight three other approaches that particularly interest us. The first two deny that a computational framework is even useful for explaining cognition. According to strong (e.g., P. M. Churchland 1981, 1982; P. S. Churchland 1983, 1986) and "ruthless" (e.g., Bickle 2003, 2006) versions of reductionism, computational and other functional descriptions of systems are at best heuristic, and genuine explanations of cognition will be purely neuroscientific. These forms of reductionism are often taken to be eliminativist about cognitive processes. On the other end of the spectrum, embodied and dynamical approaches deny that explanations of cognitive and psychological behaviors are internal processes in the ways that both computational and neuroreductionist approaches presuppose. Instead, dynamical and embodied approaches focus on dynamic patterns of interaction between organisms' bodies and brains with their environments, which they take to be fully explained by differential equations that make no assumptions about representations or other computations (e.g., Turvey 1977, Gibson 1979, Beer 1990, Kelso 1995, Chemero 2009; see also Noë 2005, Varela et al. 1991; see Shapiro 2011b for discussion). In particular, they do not explain cognition as a process involving the manipulation of realizer-independent vehicles of the sort that many philosophers and cognitive scientists take to be distinctive of computational processes (e.g., Garson 2003, Miłkowski 2013a, Orlandi 2014, Piccinini 2015). For example, Marr's theory of vision models visual perception as the process of calculating two- and three-dimensional information about the environment from the two-dimensional array of light that stimulates the retina (1982). But a number of contemporary alternative contenders do not assume that vision is a computational process designed to solve problems in "inverse optics" in order to "recover" information about the world.[5] Rather, they see vision as directly associating actions (Churchland et al. 1994) or percepts (Purves and Lotto 2003) to the world based on causal or statistical connections alone. And even some who grant that vision involves information processing deny that it is the fancy kind of information processing that makes use of computations over representations (e.g., Orlandi 2014, Purves and Lotto 2003). Of course, there remain live explanatory and empirical questions here, and we don't want to be hasty. By the same token, the advocate of multiple realizability cannot afford to be cavalier.

A third alternative to computational theories of cognition admits the usefulness of computational explanations in psychology but emphasizes their incompleteness; such explanations offer mechanism sketches or schema that do not fully explain until their mechanistic details are filled in (Piccinini and Craver 2011, Kaplan and Craver 2011, Miłkowski 2013b, Piccinini 2015; see also Barrett 2014). On this view, explanations in the cognitive sciences can be given a computational gloss, but complete explanations require mechanistic details in ways that the computational theory says are unnecessary or impossible.[6] If computational explanations in the psychological sciences are mechanistic, then they do not conceive of cognitive processes as essentially abstract, relating

---

[5] And in fact these "new" approaches trace back quite a while to the ecological psychology of Gibson (1950, 1966, 1979).

[6] Piccinini argues both that computational explanations are mechanism sketches and that brains are computers. But we see a tension in this combination: Mechanistic explanations are not complete until their details are filled in, yet Piccinini holds that computations are realizer independent. If computational explanations are realizer independent, then it seems that they are not mechanism sketches, i.e., their physical details do not need to be specified. Piccinini is aware of this concern, and he argues that we are too demanding of what counts as a mechanistic explanation; but Miłkowski (2013b) appears to accept this interpretation of the mechanistic view of computational explanation, according to which the computational explanations are incomplete until the mechanistic details are added.

For resistance to the assimilation of all cognitive science explanation to mechanistic explanation, see Shapiro (forthcoming), Barrett (2014), and Weiskopf (forthcoming).

essentially functional states. Rather, they abstractly describe concrete causal processes that occur in brains.

No doubt reductionist, embodied/dynamical, and mechanistic explanations of cognition do not exhaust the list of alternatives to the computational theory of mind. And although we don't embrace any of these three options as a universal theory, we do think that they offer serious challenges to the idea that, when explaining cognition, there is only one game in town—viz., representationalist computationalism. Thus, we feel safe in rejecting the argument that psychological states must be multiply realizable on the grounds that multiple realizability follows from the robust representationalist version of the computational theory of mind, which in turn constitutes the universal framework for psychological explanation. It is not. The abstractness and ipso facto multiple realizability of cognitive and psychological kinds is not implied by the "fact" that they are universally kinds that figure in symbol-manipulation explanations.

Another candidate for the special and ontologically significant kind of abstractness of computational descriptions or models is that they characterize their targets as information-processing systems. Now the notion of information is no less vexed than that of computation (cf. Miłkowski 2013a, Piccinini 2015). But on some notions of information processing, everything can be described as an information processor. Accordingly, if all there is to *being* a computational system is *having some description* as an information-processing system, then all cognitive processes are computational in the sense of information processing. On this view, computers are everywhere—and have been since the dawn of time (cf. Putnam 1988, Searle 1992, Chalmers 1996, Miłkowski 2013a). Some philosophers and computer scientists regard this pancomputationalist consequence as a *reductio ad absurdum* of the description-based theory of computation (e.g., Piccinini 2015). We need not take a stand on whether pancomputationalist consequences disqualify a theory of computation; sufficient for our purposes is that it fails as a theory of computation in psychology. There is no reason to suppose that cognitive scientists, when offering an information-processing model of some psychological process, would always endorse the inference that anything that can be so described or modeled is also a genuine instance of that same psychological process. The goal of information-processing explanation in psychology is generally to explain what a system does, not to describe its essence.

Remember: We don't deny that computational explanations, including information-processing and even symbol-manipulation explanations, are common in the cognitive sciences—even if we have highlighted some viable competitors. But here the emphasis is on whether computational models should be interpreted as ontologically committing. Perhaps some strongly representationalist explanations and models may be interpreted realistically, but these certainly do not constitute the only game in town. And many computational models of minds qualify as abstract only in the sense that they leave out details, or they focus on information processing. These models do not mandate the inference that such processes may themselves be realized in anything at all. It is uncharitable for philosophers to take the fact that cognitive science describes cognition in terms of information-processing tasks as implying that any system that can be similarly described should be identified as a cognitive system.[7] That rivers or digestive systems can be modeled as carrying out information-processing tasks does not show that there is river or digestive cognition, the current trend toward taking seriously plant cognition notwithstanding (cf. Calvo Garzón and Keijzer, 2009).

Some cognitive processes may be genuinely computational—information processing, and even symbol manipulating—in ways that imply their multiple realizability. But that is compatible with other kinds of cognitive processes being importantly identifiable with kinds of brain processes, and that is all that our modest identity theory requires (see also Shagrir 1998).

3 Neural Computation and Medium Independence

We've raised some doubts about the prominence and ontological portent of representational and information-processing explanations in the cognitive sciences. But we also have some doubts about whether all so-called computational explanations and models in the cognitive sciences abstract away from neural details at all. As Piccinini and Scarantino note, "In many quarters, especially neuroscientific ones, the term 'computation' is used, more or less, for whatever internal processes explain

---

[7] Robert Cummins (2000) is sensitive to different explanatory goals in the cognitive and brain sciences, though his analysis hews more closely to Marr's than we would advocate.

cognition" (2010: 244).[8] And they think that as a matter of fact the "whatever" in "whatever internal processes explain cognition" actually refers to "processing" of neural spike trains: "In recent decades, many neuroscientists have started using the term 'computation' for the processing of neuronal spike trains (i.e. sequences of spikes produced by neurons in real time). The processing of neuronal spike trains by neural systems is often called 'neural computation'" (Piccinini and Scarantino 2010: 239). Piccinini and Scarantino argue, moreover, that neural computation is not just a flight of fancy, but is, in fact, a legitimate and *sui generis* kind of computation, commenting that "such a theory need not rely on a previously existing and independently defined notion of computation, such as 'digital computation' or even 'analog computation' in its most straightforward sense" (2010: 244).[9]

Convincing Piccinini and his collaborators that "neural computation" is a kind of computation is that it involves manipulation of medium-independent "vehicles":

current evidence suggests that the vehicles of neural processes are neuronal spikes and that the functionally relevant aspects of neural processes are medium-independent aspects of the spikes—primarily, spike rates (as opposed to any more concrete properties of the spikes). Thus, spike trains appear to be another case of medium-independent vehicle, in which case they qualify as proper vehicles for generic computations. (Piccinini and Scarantino 2010: 239)

---

[8] Elsewhere they write, "Many neuroscientists use 'computation' and 'information processing' interchangeably. What they generally mean by 'information processing', we submit, is the processing of natural (semantic) information carried by neural responses" (2010: 243). See Shagrir (2006, 2012), Aizawa (2010), and Eliasmith (2010) on the various notions of computation in the cognitive and brain sciences, and Cao (2012) on an alternative account of neural information processing.

[9] We are focusing our discussion on the account developed by Gualtiero Piccinini along with several collaborators because it is the most sophisticated account of neural computation with which we are familiar (Piccinini and Scarantino 2010, Piccinini and Bahar 2013, Piccinini and Shagrir 2014, Piccinini 2015). But Piccinini's view also presents interpretive difficulties for us because he makes it an explicit criterion on any account of physical computation that commercially available digital computers and brains both count as being computers according to the theory, and formulates an account of generic computation to ensure that outcome (2015). Here we are raising questions about whether brain processes are medium-independent in the way that Piccinini says they are. But we could also ask the more fundamental question of what justifies the assumption that brains should count as computing machines.

So spike trains are medium-independent "vehicles" that are "processed" by neural computations, which themselves are a special case of generic computation. But what is "generic" computation? Piccinini and Scarantino explain:

We use "generic computation" to designate any process whose function is to manipulate medium-independent vehicles according to a rule defined over the vehicles, where a medium-independent vehicle is such that all that matters for its processing are the differences between the values of different portions of the vehicle along a relevant dimension (as opposed to more specific physical properties, such as the vehicle's material composition). Since a generic computation is medium-independent, it can be instantiated by many different physical mechanisms, so long as they possess a medium with an appropriate dimension of variation. (2010: 239)

So Piccinini and Scarantino hold that "neural computation" is a genuine kind of computation because neural spike rates are "medium independent" (Piccinini and Scarantino 2010; see also Piccinini and Bahar 2013, Piccinini and Shagrir 2014, Piccinini 2015).

In the context of thinking about the computational theory of mind and the identity theory, processing of neuronal spike trains does not initially seem like a promising candidate for being something that is abstract in a way that implies multiple realizability. Computation over neural spike rates seems to require, well, neurons. Many things can have frequencies, of course; and many things could create electrical discharges of any given frequency. There can be many causes of those effects. But, returning to a distinction we drew earlier, the frequency of the spike train of a neuron or neural assembly is part of the first-order description of neurons. It is a property of neurons as neurons, not just as implementers of some supraneural process. This suggests that spike train frequency is not medium independent in the way that would imply multiple realizability.

Recall that Piccinini and Scarantino hold the view that generic computation—that is to say, computation per se—involves operations over digits, i.e., medium-independent vehicles. And, they explain that "a medium-independent vehicle is such that all that matters for its processing are the differences between the values of different portions of the vehicle along a relevant dimension (as opposed to more specific physical properties, such as the vehicle's material composition)" (2010: 239). Thinking about this sort of medium-independence in terms of our

Official Recipe for multiple realization, it seems clear that Piccinini and Scarantino are pointing out that, for the purposes of neural explanation, only some of the properties of the neural spike trains count as relevant. So the differences among would-be realizers of the spike trains are irrelevant differences, but they share in common some features in virtue of which they can be used to explain cognition, e.g., frequency. This is clear in Piccinini's subsequent explanation of the medium-independence of computations:

When we define concrete computations and the vehicles that they manipulate, we need not consider all of their specific physical properties. We may consider only the properties that are relevant to the computation, according to the rules that define the computation . . . computational descriptions of concrete physical systems are sufficiently abstract as to be medium-independent. In other words, a vehicle is medium-independent just in case the rule (i.e., the input-output map) that defines a computation is sensitive only to differences between portions (i.e., spatiotemporal parts) of the vehicles along specific dimensions of variation—it is insensitive to any other physical properties of the vehicles. (2015: 122)

Similarly, Piccinini and Bahar write: "Computation in the generic sense is the processing of vehicles (defined as entities or variables that can change state) in accordance with rules that are sensitive to certain vehicle properties and, specifically, to differences between different portions (i.e., spatiotemporal parts) of the vehicles" (2013: 458). Rules are mappings from inputs to outputs; they do not have to be represented in the system. And, they claim, "vehicles are medium independent if and only if the rule (i.e., the input-output map) that defines a computation is sensitive only to differences between portions of the vehicles along specific dimensions of variation—it is insensitive to any more concrete physical properties of the vehicles" (Piccinini and Bahar 2013: 458).[10] But this is a strange way of talking about abstractness and concreteness. Just because the rule is sensitive to only some of the properties of the vehicles does not make the relevant properties of the vehicles abstract in a way that contrasts with their concreteness. It is plain that medium-independence, in the sense specified in this explanation of computation, does not imply that

---

[10] Piccinini and Bahar continue: "Put yet another way, the rules are functions of state variables associated with a set of functionally relevant degrees of freedom, which can be implemented differently in different physical media" (2013: 458). We are not convinced that this is just another way of making the point that the rules are sensitive to only some of the spatiotemporal parts of the vehicles. See Wilson (2010) regarding degrees of freedom.

computing systems are multiply realizable according to our Official Recipe. As described above, generic computational rules—and, *ipso facto*, computational explanations—specify some (but usually not all) of the concrete spatio-temporal properties of would-be realizers. So the physical parts of the realizers that are not mentioned in the explanation—and therefore that might vary between systems, giving the appearance of multiple realization—are irrelevant. The realizers are not differently the same, after all. They are *samely the same*.

So we are not convinced that neurocomputational explanations are medium-independent in a way that implies widespread multiple realization.[11] As with some symbol manipulation and information-processing explanations, the abstractness of neurocomputational explanations seems to be a feature of our explanatory models rather than a fact about the processes being modeled. As characterized by Piccinini, Scarantino, and Bahar, neurocomputational explanations look like examples of what Weisberg calls *minimalist idealization*: The computational models abstract away from various details of the phenomenon in order to highlight "only the core causal factors which give rise to a phenomenon" (Weisberg 2007: 642; see also Potochnik forthcoming). We could think of these models, following Piccinini and Craver (2011), as providing only sketches of explanations rather than full explanations.[12] Or we could think of computational explanations as working at lower degrees of freedom or lower energy levels than explanations that specify neural features directly (Wilson 2010, Lamb 2015). On all of these interpretations, the apparent medium-independence of computational explanations owes to the fact that they model or describe their phenomena in topic-neutral or abstract ways rather than to the abstractness or multiple realizability of their objects.

[11] If we're right, then Piccinini is wrong to say that "[m]edium-independence is a stronger condition than multiple realizability" (2015: 122).

[12] Chirimuuta (2014) argues that computational explanations are not Weisberg-style minimal models of mechanisms, nor mechanism sketches, but instead "interpretive" models that play a role in a style of a non-causal explanation that she calls *efficient coding explanation*. These models "ignore biophysical specifics in order to describe the information-processing capacity of a neuron or neuronal population. They figure in computational or information-theoretic explanations of why the neurons should behave in ways described by the model" (Chirimuuta 2014: 143). This is consonant with our view that many computational models are not intended to be ontologically committing. They explain what a system does or how it came to be, not what it "is" in an ontological sense.

The question we've set out to answer in this chapter is whether the widespread use of computational explanations in the cognitive sciences provides a reason to think that mental processes are multiply realizable in a way that rules out substantial and important identities between psychological kinds and brain kinds. That consequence does not follow from the mere fact that psychological explanations are abstract or computational in the sense that they do not mention brain processes, or that they do mention information processing. It might follow from the robust computationalist claim that psychological states or processes are themselves abstract, that is, that there is nothing to being a psychological process of a certain sort than being a computational process. But we don't think that cognitive scientists generally take themselves to be making metaphysical or ontological claims of this sort, at all. Furthermore, even if they sometimes do—and surely some do—that is not enough to falsify our modest identity theory. To undermine our view the computationalist would have to show that the universality or near universality of computational explanations in the cognitive sciences entails the wholesale unavailability of interesting and explanatory kind identifications between the cognitive and brain sciences. That was supposed to be the upshot of the "only game in town" argument. Yet once we recognize that many apparently computational explanations are not to be interpreted as implying an abstract computationalist ontology, the connection between the practice of offering computational explanations and multiple realizability is significantly weakened.

4 Methodological Considerations in Favor of Identity

In this chapter and the last, we examined indirect lines of evidence for multiple realization that involved appeals to neural plasticity, convergent evolution, artificial intelligence, and the computational theory of mind. These kinds of evidence are indirect in the sense that they do not purport to reveal actual cases of multiple realization, as did the examples we examined in Chapters 5 and 6. Rather, in the first three cases, they offer reasons to believe that multiple realization has a higher likelihood than identity even if we have not observed as much multiple realization here on Earth as some philosophers had expected. Observations of plasticity, convergent evolution, and artificial intelligence, the argument goes, are just what we should expect given multiple realization, but come as a surprise were the identity theory true. In the case of the computational theory of mind discussed in this chapter, the argument moves in the other direction. We should expect to find multiple realization because cognition is computational: The computational nature of cognition precludes the possibility of identities between psychological and neural states. We raised serious doubts about the cogency of all of this indirect evidence, and we concluded that the case for widespread multiple realizability is not compelling.

At this point, the verdict should be clear: Multiple realization of cognitive processes by brain (and non-brain) processes may sometimes occur, but evidence for its abundance is much weaker than typically assumed.

Many philosophers will find this conclusion surprising. A dogma of contemporary philosophy of mind and philosophy of science is that multiple realization is obvious, omnipresent, and inevitable. As we have been at pains to point out, multiple realization is the keystone of the standard way of understanding not only minds, but the existence and justification of all non-fundamental or "special" sciences.

But our arguments suggest that multiple realization may be more like the exception than the rule. One reason, discussed in Chapter 7, is that minds are complicated and therefore highly constrained systems. Indeed, for complex evolved traits, constraints on evolutionary options make it unlikely that we will actually observe radically different realizers even for things that are multiply realizable (Shapiro 2004). We argued that eyes are multiply realized; but also that only a handful of different kinds of eyes have actually evolved, and they share many common features and are made of the same building blocks. Given the optical constraints on eyes and the materials available to natural selection for building eyes, the presence of relatively few different sorts is hardly surprising.

Second, the brain sciences have not developed in isolation from the cognitive and psychological sciences. Rather, as Patricia Churchland has argued for many years, they have coevolved (1986; see also Kobes 1991). The example of theories of memory, discussed in Chapter 6, illustrates coevolution nicely. The cognitive psychologist offers a theory of memory that the neuroscientist refines with observations about dissociations. The structure of memory reveals itself through a combination of evidence involving psychological experimentation and neuroscientific manipulations. Thus, a theory of memory develops out of the close interplay of psychological and neuroscientific progress.

We noted that William Bechtel and Jennifer Mundale (1999) make the case that the appearance of multiple realization is sometimes created by a failure to match the "grain" of cognitive and brain explanations— cognitive explanations tending to be more coarse-grained or general, and neuroscientific explanations tending to be more fine-grained or specific. They argue that research in the neurosciences presupposes that many cognitive processes can be localized in individual subjects and in species, allowing for their neural identification across diverse species. Indeed, Bechtel and Mundale draw our attention to the simple fact that most work in neuroscience is performed on non-human and indeed non-mammalian subjects, with the background assumption being that such work can help us to understand human brains and cognition (Bechtel and Mundale 1999, Mundale 2002, Couch 2009). Whether Bechtel and Mundale are right about grain mismatches being a general source of the appearance of multiple realization, or about the degree to which coevolution of the cognitive and brain sciences depends on kind identities, it is enough for our purposes that these factors sometimes occur and can sometimes produce an illusion of multiple realizability.[13]

This integration of psychological and neuroscientific investigation makes likely that the taxonomies of the cognitive and brain sciences will frequently be in alignment. Indeed, some theorists take the strong view that the identification of psychological and brain processes is the working hypothesis for these sciences. Bechtel and McCauley, for example, defend what they call the "heuristic identity theory," according to which "Identity claims typically play a *heuristic* role in science. Scientists adopt them as hypotheses in the course of empirical investigation to guide subsequent inquiry—rather than settling on them merely as the results of such inquiry" (1999: 67). Having postulated identities as a research heuristic, it is unsurprising that experimental results often confirm the heuristic hypotheses. Bechtel and McCauley say "mapping at least some mental states (viz., many that figure in scientific psychology) one-to-one with physical states is a perfectly normal part of research in cognitive

---

[13] Bechtel and Mundale's (1999) influential argument has been the locus of much discussion. See, e.g., Kim (2002), Couch (2004), and Aizawa (2008), among others.

neuroscience and . . . the results often provide ample support for these hypotheses" (1999: 67).

It is worth noting that multiple realization seems most plausible—almost inevitable—when we adopt the dubious view that each science occupies its own level in a hierarchy of sciences, corresponding to a hierarchy of levels of the phenomena to be explained. On this view, normal scientific practice involves the development of distinctive explanations of the phenomena at each level in the hierarchy, each couched in proprietary vocabulary. In this picture of scientific practice, questions about the relationships between entities and explanations in different sciences (or at different levels) arise only after the fact or outside ordinary science—when we pose theoretical or philosophical questions about the sciences, or perhaps during episodes of scientific revolution. But it is quite clear that few if any sciences fit this picture, and certainly the cognitive and brain sciences do not. Normal practices in these sciences involve procedures, explanations, and theories that operate at many levels or grains, with no science confined to a single level or grain. This is science without blinders. Indeed, the very idea of levels in nature may make sense only in very local and specific contexts and we harbor strong doubts that the world cooperates to organize itself into neat layers across the board (Craver 2001, 2002, 2007; Craver and Bechtel 2007; Potochnik 2010, forthcoming; Shapiro and Polger 2012).

In fact, for many purposes, the cognitive and brain sciences are more usefully thought of as one science than as two distinct endeavors. It is somewhat artificial to ask about the relationship *between* the taxonomies of the cognitive sciences and the brain sciences, rather than to ask about the taxonomy—or taxonomies—of the cognitive-and-brain sciences. Of course, the standard picture of the relations among sciences makes this kind of question seem inevitable. But that is the picture that we are rejecting.

Finally, as we have emphasized throughout, scientists are very good at finding surprising patterns in nature. They are in the business of discovering similarities and regularities. These are the scientific kinds and generalizations, such as they are. They might not always retain all the features of a philosopher's idealized natural kinds and natural laws. For instance, the regularities scientists discover are rarely universal or exceptionless, and the kinds they identify are frequently messy. Nevertheless, they comprise the real tools of the real sciences. To suppose that the actual products of our sciences are deficient because they don't satisfy our philosophical expectations would be a mistake. If we want to understand explanation, our accounts had better make sense of the actual explanations we have. The job of multiple realization, we argued in the opening chapters, is to help explain the relationships between various actual sciences, their practices, and the things that they study. But it seems to us that the depiction of sciences in which multiple realization plays this role should itself be rejected because it relies on unjustified philosophical presuppositions about what those sciences are like.

PART III
After Multiple Realization

9

Putnam's Revenge

1 Taking Stock

In the first four chapters we argued that the importance of multiple realization rests on the job that it is supposed to do in showing how "non-reductive" theories, namely, realization theories, are more explanatorily General than "reductive" theories such as mind-brain identity theories. We argued that our Official Recipe for multiple realization correctly captures the pattern of variation between psychology and neuroscience that would serve the non-reductive aims of realization theorists. Further, we made the case that much of the support for multiple realization depends on question-begging philosophical arguments that betray the mere lip service given to the fact that multiple realization is an empirical hypothesis, to be decided on the basis of empirical evidence.

In the subsequent chapters we considered the actual evidence for multiple realization. There we examined some well-known examples, along with some lesser known cases that looked promising. We argued that the *prima facie* examples do not satisfy the requirements of our Official Recipe. We then argued that there are no general theoretical reasons to expect multiple realizability in the absence of actual examples.

Having examined the empirical evidence for multiple realization, we conclude that taxonomies—cognitive ontologies—that are compatible with mind-brain identities play a substantial and important role in the explanatory paradigms in psychology and neuroscience. Cognitive and psychological explanations do not, after all, employ kinds that are inevitably cross-classified by the brain sciences. And that's a good thing, too. Without such close coordination, producing the kinds of integrated multilevel mechanistic explanations and models that are typical of the "mosaic unity" of the mind and brain sciences would not be possible (Craver 2007).

The upshot of these considerations is that functionalist and realization-based theories of mental states are not, in fact, necessary for the accommodation of the actual generality of psychology. The evidence we've surveyed supports our contention that brain-based theories of mental states are sufficiently general to accommodate the actual range of psychological phenomena as we know it.

Putnam was wrong about how to tally the score in our scorecard (Table 1.1, in Chapter 1). The score, at least as far as generality goes, is a tie: Both are sufficiently general (Table 9.1). Of course each check mark in this scorecard represents an evaluation of the outcome of a heated philosophical or philosophical-cum-scientific debate. In every case, some would argue for the alternative judgment. The debate we have been considering up to this point concerns only Generality and whether there is evidence for the sort of multiple realizability of psychology that would prevent the identity theory from being adequately general. Holding fixed the other desiderata, the greater generality of functionalist and realization theories was supposed to be the reason that they should be preferred. But, we have argued, psychology is not General in that way: The actual practices of the cognitive and psychological sciences are compatible with mind-brain identity theories. Indeed, one can argue that realization and functionalist theories are overly general—overly *liberal*, in Block's terms (1978).

However, one might wonder, can we really maintain the Generality of the identity theory without "unchecking" some of the other features that

Table 9.1 Scorekeeping theories in the philosophy of psychology, revised

	Real	Causal	General	Explanatory
Dualism	✓		✓	
Behaviorism			✓	✓
Identity Theory	✓	✓	✓	✓
Realization	✓	✓	✓	✓

we mention in the scorecard? Perhaps reductive theories like the identity theory amount to versions of eliminativism: They imply that mental states are not Real, after all. Or, even if not eliminativist, maybe an identity theory undermines the Explanatory value of psychology by denying that psychology is a legitimate Explanatory science. We take these two concerns in order. In this chapter we explain why our account is not a version of eliminativism. In doing so we will develop some resources that will also be useful for explaining, in the next chapter, how our account preserves the Explanatory legitimacy and autonomy of psychology.

To begin, consider the line of reasoning behind our confidence that identity theories display the appropriate amount of Generality. If our arguments are correct, the objectives and practices of the cognitive and brain sciences give us reason to expect psycho-neural identities should be closer to the rule than the exception. In order to make this argument, we had to do two things. First, we offered an account of the phenomenon of multiple realization. Second, we provided examples of how apparent cases of multiple realization should be understood. We argued that by and large they can be explained—or explained away—in ways compatible with identifications between mental and brain states and processes. The strategies for accounting for apparent cases of multiple realization vary in step with the variety of phenomena in the world and in scientific practice that mimic the appearances of multiple realization.

But some philosophers will worry that the strategies that we deploy are self-defeating, particularly when they involve revising our categories of psychological processes. If we are going to claim that the best overall psychological theory identifies at least some mental processes with brain processes, then there had better be such things as mental processes, and these processes must play a role in explanations of behavior. If the sort of identity theory we envisage would replace psychological explanations with neuroscientific ones, or replace talk of mental states with that of brain states, suspicions that we have "gone" eliminativist would be justified.

Of course some philosophers and cognitive scientists have explicitly endorsed this sort of eliminativism—e.g., Patricia Churchland (1986), Paul Churchland (1981), and John Bickle (1998, 2003). But we reject this strategy. After all, we promised that our account is compatible with a kind of Explanatory autonomy for cognitive and psychological sciences,

Table 9.2 Scorekeeping theories in the philosophy of psychology, second thoughts

	Real	Causal	General	Explanatory
Dualism	✓		✓	
Behaviorism			✓	✓
Identity Theory	?	?	✓	✓
Realization	✓	✓	✓	✓

a topic to which we return in the next chapter. So the objection that we are considering is that we are accidental eliminativists; that our arguments against multiple realization push us towards eliminativism even if we don't realize that they do, or even if we did not intend that consequence. Call this idea that rejection of multiple realization undermines the ontological status of mental states *Putnam's Revenge.*

Were Putnam's Revenge a genuine concern, the scorecard we introduced in Chapter 1 would look more like Table 9.2, and identity theories would be conspicuously inferior to realization theories. Could that be the real state of play?

2 Unification, Individual Differences, and Convergence

As we've sketched it so far, Putnam's Revenge is little more than a hunch that our arguments against multiple realization are self-defeating. In responding to this worry, it pays to recap briefly the main arguments we have deployed:

Unification.   We argued that in some cases where a single psychological process appears to be multiply realized, it in fact turns out that the various ways of realizing the process have something in common after all—the differences in the realizers are of the wrong sort: They are irrelevant to a proper assessment of multiple realization. In that case, we can unify the various realizers and the appearance of multiple realization is erased. We offered this sort of treatment of the example of Sur's rewired ferrets (Sharma et al. 2000, von Melchner et al. 2000), arguing that the process was not correctly described as occurring variously in the visual and auditory cortex. Instead, we argued, closer examination reveals that basically the same mechanisms are employed in the brains of the normal and rewired ferrets (cf. Shapiro 2004, 2008).

Individual Differences. The Individual Differences strategy is a version of the Unification strategy. We argued that multiple realization occurs only when the differences in realizers contribute to the sameness of the realized process. That is, they have to be different ways of doing the same thing. When the differences manifest themselves in small differences in what is realized, then we have a case in which the differences account for the differences, not for the sameness. We argued that Ken Aizawa's example of the variations in retinal cone opsins should be understood in this way (Aizawa forthcoming, Aizawa and Gillett 2011). The individual differences among the human color vision capacities are not different ways of implementing one capacity—"normal color vision," as Aizawa calls it. Rather, they are the same insofar as their sensitivities are similar; and insofar as their sensitivities are not similar the differences are explained by the differences in cone opsins.

Convergence and Homology. Another special case of the Unification strategy is to provide evidence for evolutionary convergence or homology. Advocates of multiple realization are impressed by the fact that common organs appear in diverse creatures, such as camera eyes in human beings and octopuses. But the fact that camera eyes appear in diverse creatures does not by itself show that camera eyes have been multiply realized. In fact, we argued against the multiple realizability of camera eyes despite the fact that they appear to have independently evolved on multiple occasions and in multiple species. All of that is irrelevant to their classification among the kinds of eyes—they are all camera eyes (Land and Fernald 1992). We emphasized that often we find convergence because of widespread constraints in nature. And we argued that we cannot blithely suppose that those constraints are dispensable. On the other hand, in some cases the common features of distinct creatures trace back to their common ancestors—as Karten hypothesized is the case with the mammalian cortex and avian dorsal ventral ridge, which now seems to have been confirmed. Other times they trace back to a common ancestral trait from which the current traits independently evolved—as with the evolution of photosensitive opsins that appear in all sorts of eyes, and as may be the case with the electroreceptive capacities that independently became part of the jamming avoidance response in African and South American weakly electric fish.[1]

These first three strategies for understanding the scientific evidence that purports to support multiple realization should by now be familiar, and no doubt there are other unification strategies as well. Indeed, unification strategies have been around in at least a rough form since the very first multiple realization arguments for functionalism. Unification is not an eliminativist trap, however. These are strategies for showing that the apparent diversity of realizers is just that, an appearance. Behind the appearance are commonalities among neural realizers that allow identifications between psychological and neural kinds. But unification strategies do not deny the existence of the psychological phenomenon thought to be multiply realized. Failure of the inference to multiple realization carries with it no demand for elimination. Camera eyes do not disappear upon the discovery that they are not multiply realized across mammals and mollusks. But other strategies for explaining away prima facie examples of multiple realization are more worrisome.

3 Heuristics, Abstraction, and Idealization

In addition to unification strategies for minimizing the significance of multiple realizability in the cognitive sciences, we have appealed to two others with more recent origins. These interpretive strategies direct our attention to the practices of cognitive and brain scientists rather than to theories or explanations directly. First, consider the use of heuristics:

Heuristics. The idea that scientists assume the presence of mind-brain identities as a research-guiding heuristic has been advanced most directly by William Bechtel and Robert McCauley (1999, McCauley and Bechtel 2001, McCauley 2012). They argue that the cognitive and brain sciences assume the availability of mind-brain identities as a working hypothesis, especially in the design and execution of experiments, the synthesis of medical drugs, and the organization of

---

[1] The importance of homology and homoplasy for psychological kinds and the appearance of multiple realization is emphasized by Balari and Lorenzo (forthcoming).

research programs. Following Bechtel and McCauley, we argued that we should expect that the taxonomies of the cognitive and neurosciences will be in alignment—for indeed such alignment is the goal of the research and presupposed by many experimental designs.

We like to think that the discovery of mind-brain identities by the deployment of these heuristics has the justificatory effect of vindicating the heuristic assumptions and, as it were, discharging their provisional or heuristic status.[2] Early brain anatomists employed a similarity heuristic based on phylogenetic relations (Bechtel and Mundale 1999). The assumptions that many of the same brain structures can be found across mammalian species, and that more rudimentary forms of those brain and neural structures can be found in non-mammalian species that have more ancient common ancestors, have been largely borne out. Scientists are good at discovering surprising similarities in the world. As Bill Wimsatt notes, "The robustness of objects ... generates multiple richly articulated properties, and thereby provides most of the logical power arising from an identificatory or localizationist claim. This works surprisingly often even when some of the causal details are unknown or incorrectly specified" (2006: 454). Of course not every presumed mind-brain identity will be vindicated; but this hardly detracts from their heuristic importance in directing research.

Abstraction and Idealization. Sometimes explanation involves abstraction—the removal or deliberate neglect of some information in an explanation, model, or data set. Often the neglected information is thought to make no difference or only a negligible difference to the particular prediction, explanation, or procedure in question. Abstraction is a ubiquitous feature of the scientific enterprise. Arguably some kinds of abstraction are necessary for scientific understanding, perhaps because they are necessary for some degree of explanatory generality.

---

[2] We do not claim that Bechtel and McCauley take this view. John Bickle (2003, 2006) embraces the heuristic identities as methodological tools but argues that once the explanatory connections are made the heuristic ladder is kicked away, leaving only the neuroscientific kinds as genuinely explanatory. For more on heuristics and their relation to identification and localization strategies in the sciences, see Wimsatt (2006, 2007). Wimsatt also argues that these strategies—indeed reductive strategies in general—are not eliminativist (2006: 457).

We believe that the use of abstraction in psychological explanations and models sometimes produces the illusion of multiple realization. If a very abstract psychological taxonomy is compared to a neuroscientific taxonomy that is less abstract, it may appear that the psychological kinds are multiply realized by neuroscientific kinds, when in fact one is simply overlooking the variations in the psychological kinds. As discussed earlier, Bechtel and Mundale warn us to look out for these "grain mismatches" in the cognitive and neural sciences (1999).

Even absent a mismatch of grain, abstraction may suggest the reality of a psychological state or process when in fact no such commitment is intended. Colin Klein, for example, argues that scientists are frequently not ontologically committed to the abstractions in their theories (2008, 2013, 2014; see also Godfrey-Smith 2005). He proposes that advocates of the identity theory can argue that some cases of putative multiple realization are in fact cases of this sort, where the apparently multiply realized kind is not one to which scientists are committed. The appearance of multiple realization arises when theoretical abstractions are taken to posit unified psychological kinds. We argued that this illusion of multiple realization sometimes occurs when abstract computational explanations of cognitive processes are taken at face value.

The cognitive sciences are examples of what Peter Godfrey-Smith (2006) calls "model-based science" and Klein's point is that model-based science need not be equally ontologically committed to every component of their models. For example, although there are textbooks and chapters about *attention*, we would be surprised if one process underlay all attention phenomena. As we saw with the example of *memory*, it is likely that generic talk about attention offers a very general way of talking about a collection of various processes, perhaps with similar causes and effects. This does not prevent *attention* or *memory* from appearing in abstract explanatory models that focus on other psychological processes. But it should also not tempt us to reify the existence of one process *attention* or *memory* that is multiply realized. For one thing, the varieties of attention and memory are frequently dissociable, and that seems to show that those varieties are not different ways of doing the same thing.

Like abstraction, idealization is common in the sciences. Where abstraction is typically characterized in terms of removing or subtracting information from an explanation or model, idealization is characterized in terms of adding information—particularly information that is thought to be in some way inaccurate or false. For example, models in population biology commonly assume infinite and closed populations, while we know that all actual populations are finite and that almost all have some inflow and outflow of members (see Potochnik 2012, forthcoming). The Hardy-Weinberg model, describing the dynamics of gene frequencies across generations, assumes an infinite population of randomly inter-breeding organisms, free from migration, mutation, or natural selection. All of these assumptions are false in real populations. Nevertheless, idealized models—models containing idealizations—are useful and, moreover, we believe that at least some reveal truths about the actual and non-idealized world. Yet idealizations are, in one plain sense, not fully accurate. So one could worry that if psychological and cognitive kinds are idealizations then this constitutes a form of eliminativism. Just as there is no component of a car with which its center of gravity should be identified, so too, one might contend, commitment to the reality of mental states makes no sense in light of strategies like heuristic identifications, abstraction, and idealization. Perhaps, at best, there exist only cognitive and psychological predicates or concepts (cf. Kim 1997, David 1997).

For purposes of entertaining concerns about idealization and abstraction, consider the cognitive capacity for *perception*. Perceptual systems include visual, auditory, olfactory, gustatory, and tactile systems in humans, as well as systems sensitive to magnetic or electrical stimulation in other organisms (Keeley 2002). The exact delineation of sensory or perceptual systems remains controversial. The best way to understand the psychologist's use of terms like "perception" is to take them to be a highly abstract short-hand for referring to a grab-bag of distinct capacities that share something like a family resemblance.

Now the idea that the generic category of *perception* might not turn out to be anything more than an abstraction or idealization is not itself very worrisome. It's a bit like learning that *vehicle* is an idealized or abstract kind—as long as we can still board an airplane and fly to France, we don't care very much about whether there is really some feature shared by all and only vehicles. But if it turns out that many familiar psychological processes and trait kinds are invoked in psychological explanations only as abstractions or idealizations, then that would be more bothersome.

Of course it is not up to philosophers to determine whether psychological explanations or models involve abstract or idealized kinds. This seems to be a plain fact about explanations in psychology and other sciences, whether philosophers like it or not. But as metaphysicians of science we are concerned to understand the ontological commitments of these sciences, and therefore we can inquire about whether psychology makes widespread use of "mere" abstractions or idealizations, or whether they can be understood in some other ways. What should we do when we discover that many of the psychological kinds with which we had hoped to identify brain states are "mere" abstractions or idealizations? There are many options, but here we focus on three that we call Concede, Postulate, and Settle.

The first option is to Concede that the idealized kinds that figure in explanatory models and theories simply do not correspond to bits of the world. The entities and processes that figure in those models and theories are not those to which we should be ontologically committed. Conceding requires adopting an antirealist attitude toward scientific taxonomies, or at least the idealized portions of them. As we noted at the outset, this would be a deflationary approach to the sciences of the mind. We were hoping that the sciences of the mind would not just be useful, but also that they would largely vindicate our mental ontology and justify a good measure of psychological Realism.

On the other hand, the idea that abstract and idealized explanations should be treated in an antirealist manner has a distinguished pedigree. Advocates of such approaches typically concur that idealization is ubiquitous in the sciences. In general, the idea that scientific explanation normally trades in idealized models is associated with the so-called semantic view of theories, whose advocates almost invariably endorse scientific antirealism of just this sort (van Fraassen 1980, Cartwright 1983, Suppe 1989). We could simply be antirealists on a small scale, concerning some postulates of the cognitive sciences. After all, nobody now expects that Victorian melancholy will figure as a scientific psychological kind, and so it may go for other psychological kinds as well. But we suspect that idealized kinds are abundant and central in the cognitive sciences, just as in other sciences. All else being equal, we would prefer a view of the ontology of the cognitive sciences according to which psychological states are by and large Real. After all, if too many psychological states aren't Real, then it's hard to see how that could be a good thing for either realization theorists or identity theorists.

A second option is to Postulate abstract and idealized entities that correspond to the abstract and idealized taxonomic kinds. We could, that is, insist that if our best models and theories postulate certain kinds of things, then there must be entities of exactly the kinds characterized by our best theories and models. The view we're now considering notices that scientific models sometimes lump together things that are different and treats them as being the same, even when the differences are what we would call relevant differences. Such a model idealizes—pretends that these concrete things have something in common that they do not—and the result is an idealized kind. The worry is that if cognitive models abstract or idealize concrete things in and around us, then they would be false about those things. One solution to this concern is to infer that abstract and idealized entities must actually exist. This is the strategy of those who Postulate.

The notions of abstractness and idealization at stake in the conclusion of this reasoning are, for lack of a better term, ontological. The idea is not that abstract and idealized models are merely incomplete or inaccurate representations of concrete entities and processes—that was our starting point. The advocate of Postulate, rather, recognizing the abstractness and idealization in scientific models and explanations, takes those features of the models and explanations to be accurate representations of actual but non-concrete entities and processes—*abstracta*, which are members of a different ontological class from concrete things.

Put as such, the Postulate strategy is hard to recommend. The problem is not that it traffics in abstract and idealized entities; we are happy enough with them, as far as that goes. But we doubt whether they are in fact called for in this case. We prefer to think of abstraction and idealization as features of our models and explanations, not of the entities that they model and explain. The Postulator's line of reasoning reminds us of paradoxes that arise as artifacts of antiquated theories of reference according to which there must be an object for every referring noun and a property for every predicate. But much worse than that, for present purposes, it makes it difficult to see how explanations dealing in idealized objects can explain the behavior of the actual objects of our acquaintance. This same problem burdens the antirealist. So the Postulator's abstract and idealized entities don't seem to be earning their explanatory keep.

Despite these defects, the Postulate strategy remains popular—it is, basically, the strategy of the functionalist or realization theorist. The functionalist claims to discover that psychological kinds are multiply realized. Consequently, although each particular psychological process or state is realized in the brain, there is no concrete "physical" brain process kind with which a psychological kind can be identified. There are, however, "functional" brain state kinds with which psychological kinds can be identified. That is, finding that psychological taxonomies correspond only to ideal and abstract brain state kinds, one Postulates scientific entities—i.e., functional entities—picked out by those abstract and idealized taxonomies, and proposes to identify psychological entities with those. Not finding either behavioral or neural kinds that are exactly coextensive with psychological kinds, Putnam offers the hypothesis that psychological states are not brain states but are "another *kind* of state entirely... functional state[s] of a whole organism" (1967/1975: 433). It is this commitment to postulating the reality of distinct functional entities, different in kind from neural entities, that enables the functionalist to check the Real and Causal boxes in our scorecard.

Although we have no special grievance with abstract or idealized entities in general, we doubt that psychological processes are of that sort. One reason is metaphysical: Psychological processes are supposed to be causally efficacious, and abstract and idealized entities are poor candidates for causes. So we think it would be best if we could avoid falling back on Postulation.

Moreover, Postulation is not the only way to avoid antirealism and eliminativism, nor even the most prominent way to combine model-based scientific theorizing with our basic desire for Realism. Instead, we can Settle. By Settling, we simply accept the imperfection in our representational tools for doing science and making scientific taxonomies, recognizing that they will require us to make compromises such as abstraction and idealization. Nevertheless, as the success of the sciences strongly suggests, models and explanations that avail themselves of strategies like heuristics, abstraction, and idealization continue to be our most powerful tools for understanding and predicting the natural world. They are, as Catherine Elgin says, true enough (Elgin 2004; see also Potochnik forthcoming).

On the view we're outlining, when psychological and neuroscientific taxonomies fail to align, the reason will sometimes have to do with the presence of abstract and idealized kinds in the former. In Settling, we accept the explanatory necessity of such kinds, while denying that they demand an analysis that requires that they be "filled in" with multiply realized entities. Abstractions and idealizations are features of our ways of representing the world rather than aspects of the world itself.

Two points are worth noting. First, of course we assume that the neurosciences themselves will be full of heuristics, abstractions, and idealizations. But the current focus is on the cognitive and psychological sciences. We appealed to the presence of abstractions and idealizations in the cognitive sciences in order to explain away some putative examples of multiple realization—especially those having to do with computational models of cognition. And we appealed to heuristics to explain why we expect that taxonomies in the cognitive and brain sciences are frequently presumed to align out of practical and methodological necessity—and it is likely that these heuristics themselves depend on idealizations. The question now is whether the prevalence of heuristics, abstractions, and idealizations can help us to avoid multiple realization concerns and to defend the identity theory without undermining realism about cognitive and psychological entities and kinds.[3]

Second, although we appeal to abstractions and idealizations in the cognitive and psychological science to explain away some proffered examples of multiple realization, we are not the ones who introduce those abstractions and idealizations. Nor are we proposing the heuristic use of kind identities as an ad-hoc maneuver to vindicate our philosophical view. Abstraction, idealization, and heuristics are features of cognitive science quite apart from our strategic interest in them. Our suggestion, following Bechtel and McCauley (1999), McCauley and Bechtel (2001), and Klein (2008, 2013, 2014), is that attending to these features of the cognitive sciences explains why we find alignments between the taxonomies of the cognitive sciences and the brain sciences, and allows one to explain away some apparent misalignments of those taxonomies. But if the presence of heuristics, abstractions, and idealizations leads to a kind of eliminativism, then that's a fact about the sciences themselves rather than our appeal to them. If this is Putnam's Revenge, it will not favor the realization theorist over our identity theory but would instead imply that all Realist theories are misguided.

---

[3] We don't pretend to have provided a comprehensive discussion of the use of heuristics, abstractions, and idealizations in the sciences. Our aim is to indicate their relation to the debate over multiple realization, and more generally to note that these aspects of the sciences of the mind have often been neglected by philosophers of mind and metaphysicians of science.

4 Kind Splitting

A final strategy to which critics of multiple realization, including ourselves, frequently appeal could be construed as eliminativist:

> *Kind Splitting.* We argued that in some cases, the appearance of multiple realization should be taken as evidence that the psychological "kind" in question is not a unified kind, yet is also not merely an abstraction or idealization. Rather, the kind term is used to refer to a number of distinct kinds that should be divided. We have illustrated this previously: The "phenomenon" of memory has turned out to be an array of phenomena, each with distinctive features and frequently associated with distinct brain areas. "Memory" is, speaking more carefully, spatial memory, episodic memory, semantic memory, short-term memory, etc. (Roediger et al. 2002).

In Chapter 6 we argued for the permissibility of kind splitting. We also argued that whether psychologists should split a kind depends on the facts at hand: There is no general rule demanding that kinds be split or not. There may be cases in which philosophers, in their capacity as members of an inter-disciplinary community of cognitive scientists, could help make the case on general theoretical grounds for or against a specific instance of kind splitting, but that is not the norm. So for us the question is not usually whether to split kinds, but rather whether to expect that cognitive and brain scientists will frequently split kinds. Joseph McCaffrey (2015) argues that revising cognitive ontologies by splitting kinds occurs when the function of the realizing brain processes is "conserved" across different contexts. In our way of speaking, conserved neural mechanisms appear in different contexts but they do basically the same thing in each context, so that a function-to-structure mapping is available. In contrast, "variable role" mechanisms do different things in different contexts, so function-to-structure mappings are not available:

There is no canonical structure-functional relationship for multi-functional components in biology. Components with conserved roles perform the same role in different capacities while components with variable roles perform different roles in different capacities. Given recent research in cognitive neuroscience, this insight appears to be equally true of genes, organs, and neural systems. This suggests a "Functional Heterogeneity Hypothesis," which holds that the brain contains different kinds of multi-functional parts. According to this hypothesis, the brain exhibits a heterogeneous functional organization in which different regions are multi-functional in different ways. (McCaffrey 2015)

On McCaffrey's view, we should expect some revision of cognitive ontologies, and we should also expect some identifications of psychological and neural kinds. But where neural multifunctionality is variable there will not be any structure-function mappings, and the psychological kinds will not be split. So Kind Splitting is not a beast that will run amok and devour all of our familiar psychological kinds.

This is important if kind splitting has eliminativist consequences. Ken Aizawa calls the strategy "eliminate and split" because he equates splitting with the elimination of the kind that is replaced with multiple post-splitting kinds. He notes, "The 'elimination' portion of the strategy matters to multiple realization, since if the original 'unsplit' property were to remain, there would still be a property that is multiply realized" (Aizawa forthcoming). We think this is a harmless sort of elimination. After all, something new replaces the old kind, and there is typically an addition rather than reduction in the range of phenomena recognized and explained by the theory. In contrast, when witches were eliminated from our ontology, it was not for the sake of introducing a plethora of more specific sorts of wand-waving creatures. Again the case of memory is instructive: When *memory* was "eliminated" in favor of various forms of dissociable processes, the strategy not only enriched our understanding of memory but also added to the range of recognized memory phenomena. There is nothing sacred about *memory* that requires "it" to be one monolithic kind.

But we nevertheless appreciate the concern that if the practice of Kind Splitting were widespread then our resulting psychological science may well become unrecognizable, and would fail to capture certain entrenched intuitions about the goals of psychology. Fortunately, there is little reason to expect this outcome. Here we offer three theoretical reasons for our confidence.

The first theoretical issue, already mentioned, concerns the sanctity of current cognitive and psychological categories. This is perhaps the most familiar concern about kind-splitting strategies. When he reviewed the anthology in which Putnam's multiple realization argument was first published, David Lewis (1969) replied that even if one could not identify pain or hunger (respectively) with a single neurological kind across all species, it is still possible that pain or hunger (respectively) could be identified with one particular neural kind within each individual species. Thus, even if there could be no general reduction or identification of cognitive and psychological kinds, there can nevertheless be "local" reductions or identities. So even if no neural identification of pain per se were possible, there could still be identifications of pain in humans, pain in dogs, pain in octopuses, and so on. The species-specific relativization of psychological kinds is a version of the Kind Splitting strategy. The primary concern about this application of the strategy is that it would leave pain itself, the general phenomenon, unexplained. It would fail to answer the Socratic question of the one and the many. For it would not tell us what pain in humans, pain in dogs, and pain in octopuses have in common—about what makes them all cases of pain. Jaegwon Kim (1997) seeks to evade this concern by suggesting that *pain* is not a property at all, but a concept that we apply loosely, in just the way that our concept of a vegetable includes corn, tomatoes, carrots, and spinach. But this suggests to many that pain-feeling organisms do not share a common pain sensation. We lump them together merely in virtue of the similarity of their behavior, perhaps; but that's about all.

But notice that this concern becomes pressing only on the assumption that our familiar psychological categories are mainly correct or otherwise unrevisable. This would be the case, for example, if those categories were defined analytically, as Lewis (1970) and Sydney Shoemaker (1975) have sometimes held. But as we indicated earlier, we do not share that commitment. The phenomenon of multiple realization that concerns us occurs when the taxonomies of various sciences fail to align. For our purposes, scientific taxonomies are at stake, not ordinary categories. So it will be no embarrassment for us if scientific psychology fails to take up some or even many of our commonsense psychologistic categories. Our conjecture is that the relationship between the entities of scientific cognitive sciences and of the brain sciences is very intimate, so that the best overall theory of the mind will identify important mental and neural kinds. We take no stand on the extent to which folk and familiar psychological categories will find places in scientific psychology and cognitive science. So we simply do not have the concern that Kind Splitting psychologists will end up neglecting some "real" commonality in the world. We're content to allow the cognitive and psychological sciences to determine which psychological categories count as real.

A second and closely related worry is that Kind Splitting could undermine the generality of the cognitive and psychological sciences. This sort of concern arises because many philosophers have thought that if we permit Kind Splitting then the kinds will have to be extremely fine-grained. How fine? As Terry Horgan puts it, mental states are "radically multiply realizable, at the neurobiological level of description, even in humans; indeed, even in individual human beings; indeed, even in individual humans ... at a single moment" (Horgan 1993b: 308). But of course we expect our sciences to be more general than as to apply to only one individual, much less to only one individual at one specific time. This would, indeed, be bad. But none of the evidence that we have cited suggests a need to split kinds so finely that the generality of psychological explanation would be jeopardized. Individuals and even species, we have argued, are more alike than Horgan and others thought they would be. Phenomena like inter-species variation and neural plasticity do not provide grounds for multiple realization claims—in part because they are often better treated as cases of Individual Differences and Convergence, and those are no trouble for our view. Moreover, while we endorse some splitting of cognitive kinds in some cases, we have also used unification strategies to argue that some neural kinds are more general than they seemed to be at first. So we need not worry that Kind Splitting will rob the cognitive and brain sciences of any explanatory generality.

The third and final theoretical concern is that Kind Splitting would trivialize the resultant identifications. That concern is multiplied rather than ameliorated by our willingness to let the cognitive sciences determine the relevant psychological and cognitive process kinds. At risk is that the resulting identities would be between brain process kinds and psychological process kinds that no longer bear any resemblance to our familiar psychological kinds. As we have allowed, some adjustment around the periphery is to be expected. But the identity theorist can't claim a victory just by identifying entirely novel or foreign "psychological" kinds with brain process kinds. We think there are in fact two sorts of concerns here. One is that Kind Splitting produces arbitrary novel kinds, and the other is that Kind Splitting produces foreign kinds.

Suppose that in every case of a mismatch between psychological and neurological taxonomies, the psychological taxonomy should be adjusted to "fit" the neuroscience. In such a case, the matching taxonomies would not be evidence for identities but would rather presuppose them. We're not convinced that, even then, the identity thesis would be trivial. This would depend on whether the motivation for taxonomic alignment depended on mere theoretical preference or, instead, followed as something like an inference to the best explanation given a history of success with earlier identities. But here we are neither arguing for nor recommending mandatory Kind Splitting. We have simply pointed out that some purported examples of multiple realization turn out to illustrate fairly common varieties of kind splitting. Moreover, we are not adopting an arbitrary taxonomy just to make the identities work out. Rather, we presented evidence that psychological and neurological process taxonomies do tend to align and, moreover, sound methodological reasons exist to expect that result. So we cannot be accused of cooking up novel psychological kinds just to reach our ends.

But some philosophers might still be concerned that if the psychological kinds with which we propose to identify brain process kinds deviate significantly from the familiar psychological kinds, then the victory for the identity theorist is Pyrrhic at best. No doubt the identity theory was initially presented as a claim about familiar psychological kinds. The first thing to say here is that one should not so easily dismiss Pyrrhic victories. They are, after all, victories. If scientific psychological kinds are identified with neurological kinds but they turn out to be quite foreign from familiar psychological kinds, then this can hardly be considered a victory for any other theory of mind. It seems that the identity theory would be closest to the truth. However, we suspect that many important familiar psychological kinds will find their way into scientific psychology in one form or another. Neuroscience may dispense with memory, attention, perception, pain, and emotion, while leaving intact their recognizable successors. So we don't think that our arguments rely on reconstruing psychological kinds in foreign ways.

Putnam would have his revenge if in denying multiple realization we ended up also denying the Reality and Causal efficacy of mental processes. In this chapter we have examined a number of reasons that the supporter of multiple realization might see our various challenges to the doctrine as leading in this eliminativist direction. In each case—the use of heuristics, abstraction, idealization, in modeling psychological phenomena, and the methodological practice of kind splitting—we believe that the challenge to multiple realization does not require that we forsake the Reality of psychological kinds. But, another kind of worry looms for identity theorists. If psychological kinds are neural kinds, one might wonder at the need for psychological explanation. If, as we suppose, many explanations cite causes, how should we accommodate the fact that an identity between psychological and neural states entails an identity between psychological and neural causes? Should we think of psychological explanations and neural explanations as describing two overdetermining causes of the same event, or just one cause? If one, which one? Do we need psychological explanations in addition to neural explanations? These questions take us to the puzzle that Jaegwon Kim calls *Descartes' Revenge*—the puzzle of mental causation and the autonomy of psychology.

10

Mental Causation and the Autonomy of Psychology

1 Multiple Realization and Descartes' Revenge

Questions about the nature of minds—about the competing merits of realization theories and identity theories—are best understood as questions about theory choice and evidence. The well-known arguments in favor of realization theories all purport to show that they are more General than reductive or identity theories, and that this additional Generality is required by the evidence at hand—namely, the evidence for multiple realization. Those arguments fail because there is less evidence of multiple realization than realization theorists have supposed. Moreover, the identity theory, as we understand it, is sufficiently General.

But the importance of multiple realization is not limited to questions about the Generality of cognitive and psychological kinds. On a widely held view, multiple realization is central to the justification of the explanatory practices of the cognitive and psychological sciences, and of non-basic or "special" sciences in general. The basic argument, which we rehearsed in the opening chapters, is that multiple realization ensures that cognitive and psychological kinds cannot be identified with or otherwise reduced to brain or neuroscientific kinds, and for this reason genuine, *sui generis*, psychological entities stand in need of explanation. For example, Jerry Fodor writes:

The conventional wisdom in philosophy of mind is that "the conventional wisdom in philosophy of mind [is] that psychological states are 'multiply realized'... [and that this] fact refutes psychophysical reductionism once and for all. (Kim, 1992; p. 1.)" Despite the consensus, however, I am strongly inclined to think that psychological states are multiply realized and that this fact refutes psychophysical reductionism once and for all. As e. e. cummings says somewhere: "*nobody* looses [sic] *all* of the time." (1997: 149)

Fodor goes on to explain the upshot of the conventional wisdom:

a law or theory that figures in bona fide empirical explanations, but that is not reducible to a law or theory of physics, is ipso facto *autonomous*; and that the states whose behavior such laws or theories specify are *functional* states. (In fact, I don't know whether autonomous states are ipso facto functional. For present purposes all that matters is whether functional states are ipso facto autonomous.) So, then, the conventional wisdom in the philosophy of mind is that psychological states are functional and the laws and theories that figure in psychological explanations are autonomous. (Likewise, and for much the same reasons, for the laws, theories, etc. in other "special" (viz, nonbasic) sciences.) (1997: 149)

According to the conventional wisdom, then, the Reality of psychological states and the Explanatory legitimacy ("autonomy") of psychology are two sides of the same coin. We are entitled to think of psychological states as Real because functional psychological kinds figure in irreducible psychological explanations; and psychological explanations retain their Explanatorily legitimacy because they invoke functional kinds that cannot be identified with the kinds in any other science. Satisfaction of each one of these two desiderata on theories of the nature of minds is entangled with the satisfaction of the other. According to the conventional wisdom, the satisfaction of both is possible because multiple realization blocks the identification of psychological kinds with other scientific kinds, viz., neuroscientific kinds. Reality, Explanatory legitimacy, and multiple realization stand or fall together. It's a package deal. Moreover, multiple realization is supposed to do this work even if the Generality of psychological phenomena is not at issue. That is, even if realization theories and identity theories can account for exactly the same range of mental phenomena, the fact of multiply realized kinds is supposed to give the best justification for the existence and legitimacy of cognitive and psychological sciences and their explanatory practices.

Let us make three observations about the package deal. First, because we deny the ubiquity of multiple realization and reject the package deal, the conventional wisdom cannot be our view of the Explanatory legitimacy of cognitive and psychological sciences.[1] So we have some explaining to do.

Second, if we are right about the paucity of evidence for multiple realization then the cognitive and psychological sciences would not find their Explanatory legitimacy in what Fodor describes as the conventional wisdom. For if the cognitive and psychological sciences are supposed to receive their legitimacy from widespread multiple realization but multiple realization is not widespread, then the legitimacy of psychological explanation is unmoored. And that is bad news for realization theories.

Third, according to the conventional wisdom, multiply realized mental kinds are distinct from the realizing physical kinds. It is in virtue of their distinctness that they count as Real. But it also means that the causal powers of the realized kinds cannot be identical to those of their realizers—for they and their realizers are distinct. The question therefore arises: What causal powers, if any, can be attributed to cognitive and psychological state kinds?

We have arrived at the problem that Jaegwon Kim calls *Descartes' Revenge* (1998: 46). The problem is that even physicalists have difficulty explaining mental causation—particularly non-reductive physicalists such as realization theorists. Increasingly good reasons can be given to believe that the physical world is causally closed: Roughly, that everything that has a cause has a physical cause (Kim 1998, 2005; Papineau 2001). If so, how could it be that there are mental causes at all?

Descartes' Revenge is most recognizable in the form of Kim's *causal exclusion* argument (1989, 1998, 2005). Karen Bennett helpfully reconstructs the causal exclusion argument as the incompatibility of five independently plausible theses (2008: 281):

Distinctness: Mental properties (and perhaps events) are distinct from physical properties (or events).
Completeness: Every physical occurrence has a sufficient physical cause.
Efficacy: Mental events sometimes cause physical ones, and sometimes do so in virtue of their mental properties.

---

[1] As noted by Rosenberg (2001), though he seems to think this is an unwelcome consequence for us. It is not.

Non-overdetermination: The effects of mental causes are not sys-
tematically overdetermined; they are not on a par with the deaths of
firing squad victims.
Exclusion: No effect has more than one sufficient cause unless it is
overdetermined.

If the mental and the physical are Distinct, yet everything that has a cause
has a physical cause, then it is hard to see how any occurrence could also
have a mental cause unless it is causally overdetermined. Mentally
causing the drinking of a martini by desiring to raise one's glass to
one's lips does not seem to be a case of overdetermination, like a firing
squad or redundant assassin. But how could the mental exhibit Efficacy?
If for any effect we can be sure that it has a physical cause, what "work"
would be left for a mental cause to do? Positing a mental cause would
seem to be at best redundant and at worst superfluous. For this reason a
prominent project of physicalists has been to explain how their views can
be compatible with the causal efficacy of the mental.

Kim's favored solution is to deny that mental causes are Distinct from
physical causes. The physical and the mental can "both" cause an effect if
"they" are really one and the same. He argues that any theory of mind that
denies mind-brain identities will also have to deny the causal Efficacy of
the mental, on the assumptions of Completeness, Non-overdetermination,
and Exclusion (1989, 1998, 2005).

As identity theorists we are more than happy to deny that cognitive
and brain processes are Distinct. There is no causal exclusion problem
for the identity theorist; the causal exclusion problem arises only if the
mental and physical are distinct, and that is precisely what the identity
theorist denies. The identity theory therefore has no trouble accounting
for the Causal efficacy of the mental. The challenge, as we shall see, is in
explaining how identity can coexist with the autonomy of psychology.

That said, we are not convinced that the causal exclusion argument is
decisive even against those theories that maintain the Distinctness of the
mental and the neural. Various philosophers have offered compelling
reasons to reject one or more of the other theses that generate the
problem.[2] Presently, however, our concern is not whether realization

---

[2] See Yablo 1992; Clapp 2001; Bennett 2003, 2008; Block 2003; Sider 2003; Witmer 2003;
Bontly 2005; Kallestrup 2006; Shapiro and Sober 2007; Walter 2008; Wilson 2009; List and

theories can satisfactorily address the causal exclusion problem. Instead, our focus is on the centrality of multiple realization to the realizationist viewpoint. Remember: According to the conventional wisdom, multiple realization ensures the Distinctness and Reality of the mental. But, by this same wisdom, Multiple Realization, via Distinctness, also generates causal exclusion problems. There are no free lunches.

Now let's assume that realization theorists can vindicate the causal efficacy of the mental in a way that does not reject Distinctness. However, if we are right about the scarcity of multiple realization, then their in-principle success at avoiding the exclusion problem is for naught. So much the worse for the realizationist, but identity theorists, who must look for vindication of the cognitive sciences within a non-realizationist framework, will also face a problem. So how does the identity theory explain the Explanatory legitimacy of the cognitive and psychological sciences?

The answer is simple: We agree with the realizationists that cognitive and psychological explanations are legitimate because they often or always offer causal explanations. In particular, they are causal explanations according to difference-making accounts of causal explanation— about which we shall say a bit more in a moment. But even without the details we can state our solution in broad terms. Exclusion fears arise when one assumes that the presence of one kind of causal explanation— one that cites neural causes—competes with another kind—one that cites mental causes. However, identification does not create competing explanations. The identification of mental and neural processes does not force one to choose sides. We might explain a given phenomenon by citing causes of either sort. The cognitive sciences tend to cite psychological causes, and the identity of these causes with neural processes does nothing to strip such explanations of their legitimacy.

This is our answer, and we think it is the right one. There is no problem of mental causation for the identity theory. There is no question of causal exclusion. Descartes does not get his revenge. Psychological explanations are causal explanations.

But some philosophers will think that we are cheating, or worse. If the causal powers of mental states are not distinct from the causal powers of

Menzies 2009; Shapiro 2010, 2011a; Morris 2011a, 2014, 2015; Eronen 2012; but see also Ney 2007, 2009, 2012 and Kim 2012.

brain states, then why suppose that psychological explanation is distinct from neuroscientific explanation at all? This, after all, seemed to be Fodor's concern: The reduction of psychological explanations to neuro-scientific explanations seems to render them, along with the entities they posit, dispensable. In short, haven't we just traded the *causal exclusion* problem for an *explanatory exclusion* problem?

2 Explanatory Exclusion and the Autonomy of Psychology

Let us try to understand the source of explanatory exclusion worries. Suppose, for simplicity, that we think that a phenomenon is explained when we can show that it is to be expected given the circumstances and the laws of nature. Further, that we think of theories as conjunctions of sentences, where each conjoined sentence is the statement of a law or law-like generalization (cf. Hempel and Oppenheim 1948, Lewis 1970).[3] On this "axiomatic" view of theories (Klein 2013, 2014) the total conjunction of these law-like statements fixes the explanatory power of the theory: It can explain any phenomenon described in the consequents of any of its constituent law statements.[4] Let us suppose, additionally and contrary to fact, that we have such a theory about brains, B; and moreover that with B we can explain all of the brain phenomena that we know about. That is, we provisionally suppose the completeness of neuroscientific theory B. Finally, suppose that we also have a psychological explanation. Then, on the current assumptions, there is at least one psychological law, L.

Now then: What is the status of L?

On the axiomatic view (which, as will become apparent, we reject), either L is a conjunct of B, or not. If L is a conjunct of B, then the target phenomenon already has an explanation in theory B; no additional

---

[3] Fodor writes, "There is an implicit assumption that a science simply *is* a formulation of a set of laws. I think this assumption is implausible, but it is usually made when the unity of science is discussed, and it is neutral so far as the main argument of this paper is concerned" (1974: 144, fn.3). We are suggesting that the assumption is not neutral, and that it (or its ancillaries) play an important role in Fodor's picture.

[4] We hasten to add that we are not endorsing this view of explanation and theories, nor attributing this simplified version of the view to anyone. We are only using it to illustrate how explanatory exclusion concerns can arise.

psychological theory is needed to explain the phenomenon. But by calling L a statement of a psychological law we conversationally suggest that it is not a conjunct of B, that it is not part of a neuroscientific theory. What then? Perhaps if not a conjunct in B, L is implied by B. Or, if not in B or implied by B, perhaps L is translatable into some other statement L* that is in or implied by B. In these cases, L is dispensable because all of the phenomena that require explanation can already be explained by B. And if L is dispensable then L should be dispensed with—for adding L to B would not increase the explanatory power of B.

But what if L is not in or implied by B, and also not translatable into any statement that is in or implied by B? In that case, it may seem that there are three choices:

1. B was not complete after all. L should be added to B.
2. There is no phenomenon that L explains. L should be eliminated.
3. L explains a phenomenon that is distinct from any phenomenon explained by B. L should be retained as part of a new theory.

According to the first alternative, L constitutes an expansion of our brain science, B. We don't suppose there to be any hard and fast rules about when a theory should be expanded to accommodate new phenomena. The important thing for our purposes is that if L marks an expansion of B, then L does not become part of any theory distinct from B. B turned out to be incomplete, but new theory B* is now (we might suppose) complete. All explanation emerges from brain science.

The second possibility also describes a case where no theory other than B is warranted. L explains nothing at all. Maybe L turns out to describe an accidental correlation, or to be the result of error or wishful thinking. Again, the brain science provides all the explanation worth having.

The third possibility is the interesting one, and the one endorsed by non-reductive physicalists, e.g., realization theorists. On this option, L is not part of B because L explains something that lies outside the domain of B, and (however this is decided) not in the range of phenomena that B could be expanded to include. L explains something that B does not. To make the ontological commitments clear: L is part of a new theory that is justified because *there is a phenomenon* that L explains and that B does not explain.

So, the first and second strategies support the elimination of L whereas the third preserves L but only at the cost of denying that L and B can explain the same phenomena. We can now understand the motivation for explanatory exclusion:

Explanatory Exclusion: There cannot be two theories that explain any one phenomenon x unless one of the theories is redundant.

Explanatory exclusion ties the legitimacy of an explanation or theory to the impossibility of explaining its target phenomena in any other way. This may seem to be a necessary condition on the explanatory value of psychological regularities like L.

There may be other ways to arrive at the idea that explanations compete with and exclude one another, but the path from the axiomatic view of theories is quite direct. To claim completeness for a theory B is just to say that it contains all the sentences necessary to explain the phenomena within its domain. But then any new sentence, L, either adds nothing to B or describes a phenomenon that lies outside B's domain. L's explanatory legitimacy depends on its being ineluctable for the explanation of some phenomenon. Louise Antony and Joe Levine, for example, write, "a property is real (or autonomous) just in case it is *essentially* invoked in the characterization of a regularity" (1997: 91). Antony and Levine require that the regularity in question be empirical rather than analytic—the reality of corkscrews is not redeemed by the regularity that corkscrews tend to have the capacity to remove corks from bottles, for this regularity just repeats the functional specification of corkscrews. But the important part is the exclusivity created by requiring that some entity (a property, in their version) be essentially "invoked" in the characterization of the regularity. That is, there can be no characterization of the regularity that does not mention that property, and *ipso facto* there can be no explanation of the phenomenon that does not mention that property. This is what we call the *Essential Autonomy* view:

Essential Autonomy: An entity/property x is real if and only if x is essentially involved in (the explanation of) some regularity G.

Advocates of Essential Autonomy are not usually so direct as Antony and Levine. More often the commitment is implied by what they say about special sciences explanations. Putnam flirts with Essential Autonomy when he says of the example of the square peg and the round hole that the microphysical explanation is "not an explanation" (1975: 296), before settling on the weaker claim that it is inferior to the macrophysical

explanation. Daniel Dennett implies something close to Essential Autonomy when he claims that creatures that viewed the world only in
terms of microphysical entities and regularities would be "missing" something in their description of our world. If they did not also describe us as
psychological beings—Dennett says *intentional systems*—then "they
would be missing something perfectly objective: the patterns of human
behavior that are describable from the intentional stance, and only from
that stance, and that support generalizations and predictions" (1987: 25).

Philip Kitcher makes a parallel claim about the autonomy of chromosomal explanations from molecular-biological explanations of the regularity known as Mendel's second law—the genetic law of independent
assortment. Worrying that it would not be sufficiently objective to
defend chromosomal explanations on the grounds that they capture
relevant generalizations, Kitcher goes on to claim that Mendel's second
law has no cytological explanation at all:

> The molecular account objectively fails to explain because it *cannot* bring out that
> feature of the situation which is highlighted in the cytological story. It *cannot*
> show us that genes are transmitted in the ways that we find them to be. (1984:
> 350, emphasis added)

Kitcher's idea, we submit, is basically that of Essential Autonomy.
According to Kitcher there is some real feature of the genetic assortment
that can be explained only by talking about certain entities and properties, viz., chromosomes, and *ipso facto* cannot be captured by any
explanation that does not refer to chromosomes.[5]

Patricia Kitcher agrees with us that multiple realization and "irreducibility" concern scientific taxonomies, and that the "burden of the autonomy
of psychology doctrine is that psychological categories will not coincide
with neurophysiological categories" (1980: 134). But she is confident that
these taxonomic categories cannot coincide: "I think the claim that psychology is irreducible to neurophysiology rests to a large extent on the fact
that, even in principle, neurophysiology is not capable of carrying out the
explanatory work of a functional psychology" (1980: 140).[6]

---

[5] The quoted passage is immediately preceded by a footnote that compares this example
to Putnam's claims about square pegs and round holes, and alludes to considerations of
multiple realization and realizability.

[6] Kitcher is confident of this because, in the context of the paper from which we're
quoting, she assumes that psychology is purely functional. Here we are focusing on what is

And Louise Antony herself makes a similar point, writing, "there are causal regularities that *cannot be apprehended* at more basic levels of description, that require higher-level properties (like 'red blood cell,' 'hurricane,' 'wanting a cup of coffee') and that involve events which are *essentially tied to the higher level of description*" (1995: 441, original emphasis removed, emphasis added). She is even willing to extend this contention to plainly physical properties of macroscopic things: "It is highly significant that physics has no way of representing in a concise and intelligible way what—for example—diamonds have in common" (2008: 174).[7]

So the idea of Essential Autonomy is that psychological regularities are invisible and "cannot be apprehended" from the point of view of non-psychological sciences. And of course invisible things are not just those that haven't been observed, but those that cannot be observed. If Essential Autonomy is true, psychological regularities will be invisible to other sciences.

Essential Autonomy entails Explanatory Exclusion. The idea is that for any real entity, there exists some explanation in which it figures essentially, which is to say that no other explanation for the phenomenon in question exists. If there were two explanations of a regularity G, at least one of them must be wrong.

required by traditional accounts of the autonomy of psychology, and ignoring the fact that some advocates believe these criteria are satisfied because they are already convinced that psychology (or the special sciences in general) deal in functional kinds or properties.

[7] The sense in which physics has "no way" to explain what diamonds have in common with one another is that "If we wish to express some truth that we could ordinarily express in a higher-order vocabulary . . . we would have to rig up ungodly complicated combinations of fundamental physical facts, undoubtedly involving Boolean constructions of predicates of unbelievable complexity" (Antony 2008: 174). So it is not strictly speaking impossible for physics to explain the commonalities in diamonds, just very complicated; yet this is supposed to have ontological implications. Indeed, Antony repeatedly defends the idea that there is metaphysical significance to the kind and property terms that we happen to have or find useful. (See, in addition to the above, e.g., 2010: 97.) Notice that the idea that each science has its own proprietary kinds and kind terms is natural on the axiomatic view of explanations and theories (especially, Lewis 1970), but strange on the modeling view that we prefer.

We are being deliberately obtuse about the worry behind many Essential Autonomy claims, viz., that the complicated lower-level explanations would have to involve quantification over kinds or properties the predicates which are formed by disjunction and therefore cannot be genuinely explanatory kinds or properties. We take the question of multiple realization to concern actual taxonomic kinds and practices, not bringing with us any philosophical assumptions about what those kinds (or kind predicates) must be like.

Curiously, conventional antireductionists and conventional reductionists have generally agreed on the Explanatory Exclusion principle and on Essential Autonomy. We've explained its appeal for antireductionists. But why would reductionists also endorse it? The answer is that just as defenders of special science explanations argue that the legitimacy of an explanation hinges on showing that it cannot be reduced to some other kind of explanation, so too many reductionists hold that reduction is mandatory whenever possible. This is, after all, why multiple realization appears to be so important. According to the conventional wisdom, functionalists and reductionists agree on this much: Multiple realization blocks reduction. Either psychological explanations can be reduced to neural explanations, or else brains are psychologically irrelevant (Polger 2007b).

So as we have seen, Fodor worries that if psychological states can be identified with brain states, then psychological laws can be identified with neuroscientific laws. And he thinks that would be very bad indeed: "What is supposed to make the case for the autonomy (/unreducibility) of functional laws is that there aren't any laws about the realizer of a functional state *even if there do happen to be laws about the functional state that they realize*" (Fodor 1997: 156). Given coextensive laws about the realizers of psychological states, we could replace the functional psychological laws with neuroscientific laws about realizers. Then, paradoxically, psychological states would not figure in psychological laws, after all. So there would be no psychological states, after all—that is, no psychological state kinds. The worry is that the identity theory thereby implies eliminativism.

But we do not think that our identity theory leads to eliminativism because we deny the conventional wisdom, and with it the theses of Essential Autonomy and Explanatory Exclusion.

3 Explanation and Ontology

We agree with one part of the conventional wisdom: The Reality of mental states is connected to the fact that they can figure in causal Explanations. Thus, we admit that the reality of corkscrews and psychological states, respectively, depends on the fact that they figure in regularities that are cited in scientific explanations (granting a science of corkscrews), theories, and models. But we deny Antony and Levine's principle that in order to be real, a property (or entity) must figure *essentially* in some such regularity. That conception of the "autonomy" of psychology is very onerous. It makes psychological explanation of psychological phenomena mandatory—there can be no neuroscientific, behaviorist, information-theoretic, or ecological explanation. It's psychology or bust.

Notice, too, that Antony and Levine's demand makes it a condition on any explanation of any phenomenon that no other explanation of that phenomenon exists. If this were really a requirement on explanation, then we could never know that we had an actual explanation—for some other competing explanation might always be discovered at a later date. And this is not merely a matter of treating explanations as fallible, which of course they are. No one doubts that we could come to discover that we have made a mistake. But this overly demanding conception of "autonomy" implies a kind of skepticism about explanation: We may have explanations, but we are never justified in claiming to have them.[8]

On our view, psychological states are Real—at least as real as brain states. But we do not believe that accepting one explanation of a phenomenon should commit us to denying the existence of others. This restriction, we saw, may make sense on the axiomatic approach to theories because the axiomatic approach seems to give sense to the idea that a theory could be complete and so any additional claims would add nothing to an already complete theory. But we regard this as an artifact of the axiomatic view. If we think of theories as collections of models rather than conjunctions of sentences, claims of theory completeness become difficult to interpret and the idea that multiple models might explain the same phenomenon easier to accept.

Lest someone think that we are trading on an ambiguity in the idea of there being more than one explanation for a single phenomenon, let us be clear. We assume that Fodor, Putnam, Dennett, Kitcher, and everyone else recognizes different sorts of explanations, and that one phenomenon may have different explanations if they are of those different sorts. If we wish to explain why an eye has the features that it does, we may give an

---

[8] The problem here is similar to the problems with certain versions of pragmatism that make truths about the past depend on what happens in the future (Lynch 2004). In this case, what counts as an explanation now would depend on what explanations we can offer in the future. But that seems wrong.

explanation in terms of its history, its function, its causes, its causal powers, and so on. These will be different explanations, but we can probably distinguish them by being more explicit about the slightly different explananda in each case, or by citing other contextual information (Potochnik 2010). The Essential Autonomy criterion that Fodor, Antony, and Levine endorse is clearly meant to apply to explanations of a single explanandum: For example, explanations of a single phenomenon in terms of both neuroscience and psychology.[9] This seems to have been the thought behind Putnam's claim that the microphysical explanation of why the square peg will not go through the round hole is not an explanation at all, or is at least a terrible explanation. We reject this exclusionary way of thinking. Both microphysical and macrophysical explanations can be good ones, and both can be causal explanations (cf. Sober 1999). And we think this is what should be said about cognitive and neuroscientific explanations, as well. There may of course be various reasons to prefer one over the other; but the reasons will not be that one "explanation" fails to be an explanation at all.

How do we pull off this trick? The framework we need is an independently motivated model of explanation that justifies causal explanation in both psychology and the neurosciences, that permits pluralism about explanation, and that rejects Antony and Levine's principle of Essential Autonomy. Although many accounts of explanation might satisfy our requirements, we use Jim Woodward's interventionist account of causal explanation (e.g., 2000, 2003, 2008).

4 Intervention and Actual Autonomy

Most everyone agrees that the identity theory does not face a causal exclusion problem because the identity theory denies Distinctness. We are now concerned with the issue of Explanatory Exclusion rather than causal exclusion. Fodor famously opines:

if it isn't literally true that my wanting is causally responsible for my reaching, and my itching is causally responsible for my scratching, and my believing is causally responsible for my saying . . . If none of that is literally true, then

---

[9] We must be careful about the explananda, as well. We claim that even when one is very precise about the phenomenon to be explained, there is no reason to assume that only one explanation of each kind—causal, historical, and so on—is possible.

practically everything I believe about anything is false and it's the end of the world. (1989: 156)

And we might say the same thing about causal *explanation*, as well: If Sophia's wanting doesn't explain her reaching and Thalia's itching doesn't explain her scratching...then practically everything we believe about anything is false, at least as far as psychological explanation goes. More than just being surprising, that would undermine the argument in favor of the identity theory that we've been developing. So evading the Causal Exclusion problem will be pointless if we can't also avoid concerns about Explanatory Exclusion. Yet we claim that this second achievement is an easy one: All we need is a suitable account of causal explanation.

The account that we have in mind is a difference-making theory of causal explanation. Speaking in the vernacular of events for convenience: Explanations cite causes, and an event C causes an event E if, roughly, changes to C make a difference to E. The important feature of the view on which we depend is that C's being a difference maker for E is entirely compatible with their also being another event, D, that is also a difference maker for E. Difference-making accounts of the sort we have in mind simply have no use for the principle of Explanatory Exclusion. (They also, in our view, have no use for the principle of Causal Exclusion; and realization theorists can exploit that feature of the view to defend against causal exclusion arguments. But that is not the argument that concerns us at the moment.)

In greater detail, we favor an "interventionist" or "manipulationist" version of the difference-making theory (Woodward 2003), according to which difference making is understood in terms of actual or possible ideal interventions on C.[10] The modal force of the counterfactual claim— viz., that the intervention on C *would* make a difference to E—is backed by actual invariances relating C and E. Invariances are regularities among phenomena ("variables") in the world that may in some cases

---

[10] Some philosophers (e.g., Baumgartner 2009, 2010, 2013) argue that there are technical defects in Woodward's formulation that make it unsuitable for explaining causation in the special sciences. We think that those problems can be ameliorated; and at any rate they depend crucially on premises that are rejected by the identity theorist. But for present purposes, we need not be committed to the details of Woodward's account; what is important is the general idea of causal explanation as explanation in terms of difference makers.

Figure 10.1 A Rube Goldberg machine: The Self-operating napkin. Artwork Copyright © and ™ Rube Goldberg Inc. All Rights Reserved. RUBE GOLDBERG ® is a registered trademark of Rube Goldberg Inc. All materials used with permission. <http://www.rubegoldberg.com>.

be local and temporary. For example, an invariance may hold at a certain time because a Rube Goldberg machine has been built (Figure 10.1). We mention this feature of the interventionist account because it illustrates how our approach differs from the usual way of thinking of mental causation and causal explanation in terms of exceptionless or *ceteris paribus* laws of nature. Laws do not play a prominent role in the interventionist account of causal explanation. Nevertheless, intervening on or "wiggling" one part of a Rube Goldberg machine can make a difference to or "wiggle" another part of the machine. That is, there is an invariant relationship between the parts of a working machine so that interventions on events in one place can make a difference to events in another.

We believe that causal invariances hold between mental states or processes and physical and mental effects; and those are the basis for explanation in the mind and cognitive sciences as well as in the neurosciences (Craver 2007). For present purposes we can take this claim about invariances to be uncontroversial, as it is less demanding than the more familiar law-based approaches that many philosophers of the mind and cognitive sciences accept. Surely if there aren't any psychological invariances, then neither are there any laws of psychology and we're not concerned here with general doubts about psychological explanation.

We've already said that the interventionist version of the difference-making account of causal explanation does not require Explanatory Exclusion. Nor do interventionists accept Antony and Levine's principle of Essential Autonomy. So there is no trouble with reconciling mental causal explanation and neuroscientific causal explanation, even when mental processes are identical to neuroscientific processes. Assume for the sake of argument that explanations can be distinct even if the phenomena that they explain are not—explanations are intensional or hyperintensional. In the interventionist framework there is no obstacle to having more than one causal explanation for a phenomenon. Exclusion principles are simply not part of the view.

Consequently, we see no reason to view psychological explanation as in any sense less legitimate than neuroscientific explanation. Note first that on our account psychological explanations are causal explanations like any other causal explanations. That's the most important feature. They are causal explanations that pick out actual causal invariances in the world—actual cause explanations, we could say. Second, in the broadly physicalist framework that we presuppose, mental states are fully Real despite being ultimately dependent on the microphysical. We've already noted that identity theorists and realization theorists are all "reductionists" in that minimal sense. We all hold that mental processes depend on neural processes, and perhaps other bodily and environmental processes as well. As far as we know, for any effect of a mental process there can be a difference-making explanation that does not appeal to any distinctively mentalistic variables, vocabulary, or models. The parallel observation holds for explanations in any non-fundamental science like chemistry or biology. We do not suppose that psychological explanation is ever mandatory. But, having rejected substance dualism, we also don't see why we would require it to be.

Finally, let's consider the worry that our defense of the autonomy of psychological explanation is unacceptably tepid. According to us, psychological explanations gain their legitimacy from their identification of actual causal invariances. Nothing more is required; there is no demand for irreducibly mental kinds. On the conventional view, in contrast, the legitimacy of psychological explanations rests on their indispensability—on their capacity to explain what no other science could. In our view, this misunderstands the significance of autonomy. The autonomy of a science should not depend on the representational capacities of *other* sciences, but instead on its own power to describe, explain, and predict phenomena using the resources of its proprietary vocabulary and methodology.[11] That psychology can do this suffices for its autonomy, despite retaining a taxonomy in rough alignment with neuroscience.

Call the kind of autonomy we defend here *Actual Autonomy* because it says that psychological explanations gain their legitimacy when they explain effects by citing their actual causes (cf. Pearl 2000). Actual Autonomy is compatible with the existence of significant constraints between neuroscientific and cognitive or psychological kinds and explanations. But it holds that cognitive and psychological explanations remain autonomous in the sense that they satisfy the standard criteria on causal explanation. Fodor was right, we think, when he said that "it is often the case that *whether* the physical descriptions of the events subsumed by [special sciences] generalizations have anything in common is, in an obvious sense, entirely irrelevant to the truth of the generalizations, or to their interestingness, or to their degree of confirmation or, indeed, to any of the epistemologically important properties" (1974: 103). But Fodor and Putnam erred in thinking that this kind of epistemic autonomy for psychological explanations can occur only if psychological kinds earn their irreducibility through their multiple realization in neuroscientific kinds.[12] The cognitive sciences are Actually Autonomous from the brain sciences even with strong constraints between those sciences, including important identities between mental and brain processes. The Actual Autonomy of an explanation is, we think, a more robust sort of autonomy than Essential Autonomy, for it denies that the legitimacy of an explanation in one science can be held hostage to the taxonomic and explanatory claims of any other sciences or models.

---

[11] We take our view of the autonomy of psychology to be in the same spirit as Richardson (1979, 1982, 2008, 2009), even though he is skeptical of some details of our account of multiple realization.

[12] See also Haug (2011a, 2011b).

5 Identity and Explanation in the Mind-Brain Sciences

We opened this book with an account of four desiderata that we believe any successful psychological theory must meet. We set ourselves the challenge to show that an appropriately characterized identity theory satisfies these desiderata. In meeting this challenge, we take ourselves to have produced a theory with all the benefits typically associated with realization theories. And, insofar as we have cast doubt on multiple realizability—the doctrine central to functionalism—we regard the identity theory as the current best hope for understanding the relationship between the mind and brain sciences.

According to identity theories, mental states are Real, Causally efficacious, and appropriately General; and cognitive and psychological Explanations are legitimate (Table 10.1).

While our concern in this chapter has been so far focused mainly on addressing exclusion-inspired challenges to the Explanatory legitimacy of the identity theory, we need also to consider an objection of another sort.

When we claim that the best overall model of psychological and neuroscientific processes makes substantial and important use of identities, we take ourselves to be engaging in something like inference to the best explanation. We are not alone in thinking that identity theories can be supported by inference to the best explanation. Arguably U. T. Place (1956), J. J. C. Smart (1959), and Herbert Feigl (1958) all thought that the identity theory was justified in that way. And recently Chris Hill (1991), Ned Block and Robert Stalnaker (1999), and Brian McLaughlin (2001) have argued for versions of the identity theory using inference to the best explanation. Against this emerging consensus, Jaegwon Kim has argued that identities do not explain, but only allow us to rewrite explanations: "Identities seem best taken as mere rewrite rules in inferential contexts; they generate no explanatory connections between the explanandum and

Table 10.1 Scorekeeping theories in the philosophy of psychology, revisited

	Real	Causal	General	Explanatory
Identity Theory	✓	✓	✓	✓
Realization	✓	✓	✓	✓

the phenomena invoked in the explanans; they seem not to have explanatory efficacy of their own" (2005: 132). If he were correct, Kim would have cast doubt on an important piece in our overall defense of the identity theory. For we argued that among the reasons to endorse the identity theory is the fact that cognitive and neuroscientists rely on these identities in their ordinary explanatory practices. Below we respond to Kim's objection. In the process we can highlight the differences between our view and some familiar philosophical approaches to explanation in the cognitive sciences that we reject—views shared, we think, by both Fodor and Kim.

To begin, Kim assumes that explanatory arguments include premises that describe empirical laws or lawlike regularities.[13] Explanations, that is, must contain an "explanatory premise" (Kim 2005: 135). Moreover, Kim assumes that the "explanatory premise" must link distinct facts. For this reason he complains of identity explanations, "There is no movement here from one fact to another, something that surely must happen in a genuine explanatory argument" (2005: 132). Kim doesn't say how he is thinking about facts, but his examples tell us what we need to know: He holds that identities stated with rigid designators, including "Kripkean" or "theoretical" identity claims, do not link distinct facts. He illustrates this claim with an example involving proper names—the explanation of the fact that Cicero is wise from the fact that Tully is wise and the identity: "Cicero is identical to Tully." Kim says, "The fact represented by the first premise 'Tully is wise' is the same fact as the fact represented by the conclusion 'Cicero is wise'" (2005: 132). That is why there is no "movement" from one fact to another in identity explanations. Because identity claims don't link distinct facts, Kim thinks, they are entirely unsuitable for being the "explanatory" premises in deductive explanations. Therefore, strictly speaking there are no such things as explanations that invoke identities at all. Hence, inference to the best explanation could never give us reason to prefer an identity explanation over a correlation or realization explanation. And so, Kim concludes, the inference to the best explanation argument for the identity theory is "seriously flawed and incapable of generating any real support" (2005: 125).

---

[13] For additional concerns about Kim's conception of the role of scientific identities, and the motivations and evidence for them see Wimsatt (2006: 451–2). See also Morris (2011b).

We do not share Kim's assumptions about explanations and inference to the best explanation, but we think that many philosophers of mind and metaphysicians do. Kim's conception of explanation provides a metaphysics-friendly version of Carl Hempel's deductive-nomological model of scientific explanation. His update adds some metaphysical gravitas to the "explanatory" premises, whereas Hempel hoped to get by with formal structures (Kim 1999; cf. Hempel and Oppenheim 1948). But the general schema is familiar. Causal explanation seems to require an "explanatory" regularity that links distinct facts. Yet, the important thing to notice is that this requirement arises not from the demand that explanation be deductive, but because causation links distinct entities or events, and causal explanation characterizes causal regularities, invariances, or patterns in the world (Woodward 2003, Potochnik forthcoming).

Marc Lange describes two varieties of identity explanations, which he calls *crossidentity* and *simidentity* explanations (forthcoming). Simidentity explanations explain "why there obtains a certain similarity (hence the term '*sim*identity') between (what appear to be) two facts" (Lange forthcoming). In Kim's language, *simidentity* explanations explain correlations. *Crossidentity* explanation "explains a given fact by appealing to another fact and that the properties (or particulars) figuring in the two facts are identical" (Lange forthcoming). This sort of identity explanation uses identities as additional "explanatory" premises in causal explanations. The details of Lange's account do not concern us here, but two features of Lange's view are important. One is that although Kim and Lange both take explanations to connect "facts" it is clear that they employ different ways of thinking about facts. According to Kim, that "Cicero is wise" is the same fact as that "Tully is wise." This is a rather coarse-grained way of thinking about facts. Lange defends a much finer-grained way of thinking about facts and explanations, according to which facts about Cicero are distinct from facts about Tully, even if Cicero is identical to Tully. In framing his objections to inference to the best explanation arguments, Kim seems to assume that explanations create extensional contexts, individuated by the things ("facts") that they concern. But explanations are in part epistemic devices, so it is commonplace that explanation, like belief, is not always preserved when co-referring terms are substituted for one another. The belief that "Obama is the President" differs from the belief that "the winner of the last election is the President" even if Obama is the winner of the last election. Indeed, Lange argues that explanation is *hyperintensional*, meaning that it creates contexts in which even necessarily co-referring terms cannot be substituted: An explanation of the fact that "Cicero is wise" may not explain the fact that "Tully is wise." Lange writes:

Though "being water" and "being composed of H_2O molecules" necessarily designate the same property, they pick it out in different ways...Because "being water" and "being composed of H_2O molecules" pick out the same property in different ways, they differ in the information they supply and thus differ in their roles in causal explanations, since causal explanations work by supplying contextually relevant information about the explanandum's causal history or, more broadly, about the world's network of causal relations. (forthcoming)

Having observed that explanatory contexts are hyperintensional, Lange goes on to show the kind of explanatory work that identities can perform. This is good news for us; but it also reveals a second important feature of Lange's approach that is just as important for our purposes. Rather than puzzling over whether identity explanation is possible, Lange recognizes the ubiquity of identity explanations in the sciences and sets out to understand how they work. From this point of view, arguments that reject identity explanations must be making some mistaken assumptions about explanation. We have suggested that Kim's objections make such a mistake.

We've thus far focused on Kim's assumptions about the nature of the facts that, he believes, must be connected in explanatory premises. Another assumption that we question involves a sharp distinction he draws between scientific explanation and other "explanations." For example, he is concerned to reinterpret a version of the inference to the best explanation argument put forth by Brian McLaughlin. McLaughlin writes, "[The Type Identity Thesis] implies the Correlation Thesis...We maintain, on grounds of overall coherence and theoretical simplicity, that the explanation of the Correlation Thesis that the Type Identity Thesis offers is superior to the explanations offered by other theories of mind" (2001: 319–20). But Kim, having argued against the tenability of identity explanations, offers the following "charitable" reconstruction of McLaughlin's claim: "I think the 'explanation' McLaughlin speaks of here could be taken—and this may well be the way he takes it—to be a philosophical explanation, or philosophical theory building, rather than garden-variety scientific explanation to which something like the rule of inference to the best explanation might apply" (2005: 147). In offering this reconstruction, Kim appeals to a distinction between scientific explanation and other forms of explanation and furthermore suggests that inference to the best explanation can only be used to support the former.

We wish to resist the idea that scientific explanation is of an entirely different sort than "philosophical" explanation. We don't doubt that scientists and philosophers often offer different kinds of explanations. Philosophers and scientists also frequently engage in distinctive practices aimed at various and distinctive subjects. But we don't have any reason to presume a strict division between two sorts of explanation, scientific and philosophical. Scientists are just as interested in theory building as philosophers, after all. We regard the choice between metaphysical theories of the nature of minds as itself just an example of theory choice like any other. This is an approach that Kim finds puzzling (2005: 142), but we find quite natural. We expect that science and philosophy are both forms of inquiry, and that we and our colleagues who conduct experiments are jointly engaged in, as we quoted in our preface, trying "to understand how things in the broadest possible sense of the term hang together in the broadest possible sense of the term" (Sellars 1963: 1).

Moreover, the practices of the sciences themselves are among the things we hope to understand. The Identity Theory helps us to make sense of the relationships between explanations and theories in different sciences. These inter- or trans-scientific facts about the "coevolution" of the mind and brain sciences are facts about the world like any others. Identities play a role in understanding how the practices, experiments, methods, and results of different cognitive and brain sciences are coordinated. They also play a role in predicting future observed correlations that we would otherwise not look for (Wimsatt 2006: 455). We argued in Chapter 8 that the coordination of these practices lead us also to expect their taxonomic and explanatory kinds to be aligned.

Some readers will wonder why we take Kim's objections as seriously as we do. Like Lange, we supposed all along that mind and brain scientists do in fact identify psychological kinds with neuroscientific kinds. So this looks like a case where the philosophical theories of explanation need to catch up with the empirical science. But Kim's assumptions are not idiosyncratic; they are part of the standard view of scientific explanation in the special sciences and of the relationships between different sciences.

That view obscures how something like the identity theory could possibly be correct. We have offered an alternative framework that is compatible with the identity theory as we understand it. And by examining some of the specific points of disagreement between Kim and ourselves, we aimed to show that our rejection of the standard view is not ad hoc. Rather, the standard view rests on numerous assumptions about science and philosophy, about explanation and theory. There are good reasons to reject many of these assumptions even if you resist our conclusion, even if we're wrong about the evidence for multiple realization.

6 The Special Sciences Reconsidered

We began this book with the suggestion that questions about the nature of minds are questions about theory choice—about which theory to prefer. We set out a framework of desiderata and showed how the standard view of the special sciences that supports realization theories also crucially relies on multiple realization. The hierarchical picture of the world and the structure of sciences consistent with that view is widely shared. However, we argued, multiple realization, to fulfill its promises, cannot be just any sort of variation in the world. On the account we develop, multiple realization requires relevant differences and differences that contribute to sameness. Mere variation is not enough.

We went on to argue that the range of multiple realization is smaller than most philosophers of mind have supposed. Having sorted through the array of issues on which judgments of multiple realization must rest, the nearly universally accepted idea that multiple realization is a feature of all mental states begins, in our view, to look very suspicious. Indeed, the empirical burden of demonstrating multiple realization begins to look quite daunting. Insofar as our arguments have been correct, the multiple realizability thesis should no longer enjoy its status as the default assumption in philosophers' theories of mind. Moreover, we doubt that multiple realization has ever been the default assumption in the laboratories that supply us with evidence about the mind's workings. As we argued, the identification between kinds in the cognitive and brain sciences plays an important explanatory role in the best overall model of those sciences and their domains.

Nevertheless, we further argued, the presence of identities does not undermine the value or legitimacy of the psychological and cognitive sciences—the autonomy of the special sciences, in the sense worth valuing. In doing so, we advocated an alternative picture of the world and the sciences, one in which some kinds from some sciences are identified with some kinds from other sciences. Others are not. Some sciences are more anchored or constrained; others less so. On this view, the legitimacy of the special sciences has nothing to do with whether the variation in our world happens to be the sort that makes for multiple realization. Paradoxically, the importance of understanding multiple realization and the evidence in its favor rests on exposing it to be not as important as several generations of philosophers have supposed.

Our view advocates the replacement of an older but entrenched conception of how the special sciences relate to each other and to the more basic sciences, with a newer conception that disclaims Essential Autonomy and recognizes the importance that identity claims often play in the sciences of the mind. Capturing the older conception is the famous diagram of Fodor (1974) that we introduced in Chapter 1 (Figure 1.1).

Fodor's diagram combines the Hempelian idea that special sciences consist in law statements with the realizationist's belief that special science kinds are multiply realized in the kinds of a more basic science. Having rejected both these claims, we offer Figure 10.2 as a more useful picture of how the special sciences stand in relation to each other.

Absent from this diagram is the neatly ordered hierarchy of scientific disciplines that Oppenheim and Putnam (1958) once imagined, as well as the emphasis on laws that still dominates discussion of scientific explanation and the metaphysics of science. There is no layer cake of levels of sciences (see also Craver and Bechtel 2007; Craver 2007; Bechtel 2008; Potochnik 2010, forthcoming; Walter and Eronen 2011). The large circles might be thought of as collections of phenomena that belong together because of similarities in their properties or behavior and, accordingly, similarities in the methods and models necessary for their investigation. For instance, the phenomena in Domain A might consti-tute kinds of memory systems, or more broadly, psychological capacities. Communication between the A researchers and those who research phenomena in Domain B might reveal interesting identities. Perhaps "B scientists" focus mainly on neural events, and their study of brain-damaged patients leads them to postulate an identity between a B process and an A process. Similarly, "C scientists" might be mostly engaged in understanding biochemical processes; and their investigations have led

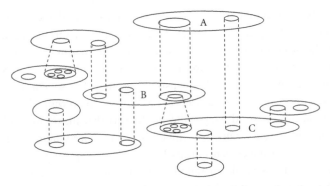

Figure 10.2 Our illustration of the relations between different explanatory theories or models emphasizes the importance of identifying some entities across different sciences, explanations, or models (e.g., A, B, C). But our picture does not assume that identification is always possible, and dispenses with the assumption that different explanations and models can be neatly ordered into layers or hierarchies.

to the interesting discovery that some A process or kind is identical to a C process or kind, but that a particular B process or kind is in fact multiply realizable in C processes or kinds. Applied to the special case of the mind and brain, Figure 10.2 is a way of illustrating our conclusion that the best overall model of psychological and brain processes makes substantial and important use of process kind identities.

Although Fodor's picture—the standard picture—rejects Oppenheim and Putnam's optimism that the sciences will be unified, it does not depart from the idea that sciences will stack neatly atop each other, with the kinds of those higher in the stack (multiply) realized in the kinds from those lower. Figure 10.2 is a much more realistic and empirically informed depiction of the relationships between the sciences. In pursuit of understanding some phenomenon, such as memory or vision or face recognition, a scientist might find enlightenment from many sources: Neuroscience, biochemistry, molecular biology, optics, information theory, and so on. The discovery of identities between scientific kinds, as we have urged in this chapter and the previous, needn't be a step toward either elimination or explanatory redundancy. Indeed, as we have urged, such identities can play an explanatory role in the best overall model of the mind and brain sciences.

We have argued that it is a mistake to think that the special sciences must resist identities—that they must proceed in isolation, proudly capable of carrying on without any help from each other, and that this is guaranteed by the multiple realizability of their kinds. And from the point of view we've elaborated in this book, it is very fortunately false. Variation in the world is everywhere, but scientists excel in finding whatever surprising similarities and regularities might weave through this variation. From the perspective we have defended in this book, it is a modest identity theory rather than a realization theory that offers the better account of entities and explanations in the cognitive and brain sciences.

Guide for Teaching and Learning

This book is intended to advance our view of the philosophical significance of and empirical evidence for multiple realization. But we took efforts to ensure that the structure and style of the book will make it accessible to non-experts, including scientists, students, and anyone who is generally curious about issues concerning the relationship between mind and brain.

Below we provide some questions to guide reading and discussion, including some questions about the soundness of our own arguments. And we direct the reader to some of the most important works with which we engage.

General Background

Four classic articles lay out the views that are the *loci classici* for the philosophical importance of the ideas discussed in this book:

Block, N. and J. Fodor. 1972. What Psychological States Are Not. *Philosophical Review* 81: 159–81.

Fodor, J. 1974. Special Sciences, or the Disunity of Science as a Working Hypothesis. *Synthese* 28: 97–115.

Putnam, H. 1967. Psychological Predicates. Reprinted as "The Nature of Mental States" in Putnam 1975.

Putnam, H. 1973. Philosophy and Our Mental Life. Reprinted in Putnam 1975.

We discuss these texts in almost every chapter of the book. Another important text in this canon, but less prominent in this book, is:

Kitcher, P. 1984. 1953 and All That: A Tale of Two Sciences. *Philosophical Review* 93 (3): 335–73.

Chapter 1

1. What broad considerations might favor a physicalist theory of minds over one like Descartes', which views minds as a special sort of non-physical substance?

2. Scientific theories are often evaluated on the basis of various virtues, such as generality, empirical adequacy, and simplicity. Why did early functionalists like Putnam believe that their theory of mind displayed these virtues to a greater extent than did the mind-brain identity?

3. Polger and Shapiro claim that successful theories of psychology require that mental states be Real, Causal, Explanatory, and General. Describe these features of mental states, and explain how multiple realizability is supposed to provide these features. Are there other features that successful theories of psychology should account for, as well?

Additional reading

Fodor, J. 1974. Special Sciences, or the Disunity of Science as a Working Hypothesis. *Synthese* 28: 97–115.

Kim, J. 1998. *Mind in a Physical World*. Cambridge, MA: MIT Press: Ch. 1.

Putnam, H. 1967. Psychological Predicates. Reprinted as "The Nature of Mental States" in Putnam 1975.

Smart J. J. C. 1959. Sensations and Brain Processes. *Philosophical Review* LXVIII: 141–56.

Chapter 2

1. What does it mean to say that realization is a synchronic ontological dependence relation? Provide some examples of kinds that are realized and of the kinds that realize them.

2. Consider the definition Polger and Shapiro offer of realization (p. 22). Is water a kind that is realized? Hydrogen? Why or why not?

3. How do so-called first-order properties differ from so-called second-order properties, and how do realization theories make use of this distinction?

4. Some kinds are realized and some are not. Provide examples of both realized and non-realized kinds and justify why you place them into these categories.

Additional reading

Bennett, K. 2011. Construction Area (No Hard Hat Required). *Philosophical Studies* 154 (1): 79–104.

Polger, T. 2007. Realization and the Metaphysics of Mind. *Australasian Journal of Philosophy* 85 (2): 233–59.

Polger, T. 2013. Realization and Multiple Realization, Chicken and Egg. *European Journal of Philosophy* 23 (4): 862–77.

Polger, T. and L. Shapiro. 2008. Understanding the Dimensions of Realization. *Journal of Philosophy* CV (4): 213–22.

Putnam, H. 1967. Psychological Predicates. Reprinted as "The Nature of Mental States" in Putnam 1975.

Chapter 3

1. What argument might justify Putnam's belief that distinct kinds of organisms more likely share a particular functional organization than a particular physical organization?

2. Putnam thought that the eyes of mollusks and vertebrates did not illustrate a case of multiple realization even though they may differ in some physical respects. What considerations justify this assessment?

3. When organisms with indisputably different kinds of brains are capable of performing the same kind of behavior (e.g. face recognition) a question arises whether their respective psychologies are the same. Explain why this question is important for assessing the possibility of multiple realization.

4. We've all seen science fiction movies that present robots or organisms with human-like minds despite lacking anything that resembles a human-like brain. What does our ability to imagine such things tell us about the possibility of multiple realization?

Additional reading

Block, N. 1978. Troubles with Functionalism. C. W. Savage (ed.), *Minnesota Studies in the Philosophy of Science*, Vol. IX. Minneapolis, MN: University of Minnesota Press. Reprinted in Block 1980a.

Gillett, C. 2003. The Metaphysics of Realization, Multiple Realizability and the Special Sciences. *Journal of Philosophy* 100 (11): 91–603.

Lewis, D. 1980. "Mad Pain and Martian Pain." In Block 1980a.

Shapiro, L. 2000. Multiple Realizations. *Journal of Philosophy* 97 (12): 635–54.

Chapter 4

1. Explain and motivate each of the four criteria in Polger and Shapiro's Official Recipe for multiple realization. What does it mean for two objects to be "samely different but differently the same"?

2. Consider Polger and Shapiro's discussion of the three watches, two of which are mechanical and the third quartz, but all of which tell the time a bit differently. How do the four conditions in the official recipe justify the verdict that the mechanical watches are the same kind of realizers of a watch, whereas the quartz watch is a different kind of realizer?

3. Develop an example of your own involving an artifact that is analogous to the watch discussion.

Additional reading

Millikan, R. 1999. Historical Kinds and the "Special Sciences." *Philosophical Studies* 95: 45–65.

Polger, T. 2009a. Evaluating the Evidence for Multiple Realization. *Synthese* 167 (3): 457–72.

Shapiro, L. 2000. Multiple Realizations. *Journal of Philosophy* 97 (12): 635–54.

Shapiro, L. 2008. How to Test for Multiple Realization. *Philosophy of Science* 75 (5): 514–25.

Shapiro, L. and T. Polger. 2012. Identity, Variability, and Multiple Realizability. In S. Gozzano and C. Hill (eds), *The Mental and the Physical: New Perspectives on Type Identity*. Cambridge: Cambridge University Press.

Chapter 5

1. What is the point of the discussion of *situs inversus*? Describe some reasons that the hearts of normal and situs inversus individuals might, after all, count as different realizations.

2. An interesting morphological dimorphism in human beings is the sexual dimorphism of males and females. Are men and women multiple realizers of the kind human being according to Polger and Shapiro? Should they count as multiple realizers of human beings?

3. The argument that vision in the rewired ferrets does not constitute a case of multiple realization makes use of the same kind of reasoning that Polger and Shapiro employed in the discussion of the watches in Chapter 4. State the reasoning in an abstract form in order to make the similarities explicit.

4. Polger and Shapiro argue that neural reuse or degeneracy does not entail that mental states are multiply realized or realizable. Why do they think so? Could reuse or degeneracy be a problem for their identity theory, even if it does not imply that mental states are multiply realized?

Additional reading

Anderson, M. 2010. Neural Reuse: A Fundamental Organizational Principle of the Brain. *Behavioral and Brain Sciences* 33: 245–313.

Barrett, D. 2013. Multiple Realizability, Identity Theory, and the Gradual Reorganization Principle. *British Journal for the Philosophy of Science* 64 (2): 325–46.

Bechtel, W. and J. Mundale. 1999. Multiple Realizability Revisited: Linking Cognitive and Neural States. *Philosophy of Science* 66: 175–207.

Block, N. and J. Fodor. 1972. What Psychological States Are Not. *Philosophical Review* 81: 159–81.

Sharma, J., A. Angelucci, and M. Sur. 2000. Induction of Visual Orientation Modules in Auditory Cortex. *Nature* 404: 841–7.

Chapter 6

1. What is the "kind-splitting" strategy and how does it challenge claims of multiple realization? What neuroscientific evidence might be used to justify kind splitting?

2. Why does Aizawa and Gillett's cone opsin example fail the fourth criterion in Polger and Shapiro's analysis of multiple realization?

3. Consider Keeley's discussion of the multiple realization of the JAR mechanism in electric fish. Polger and Shapiro argue that the evidence to date leaves open the question whether the JAR mechanism is in fact multiply realized. Describe evidence that would tilt the verdict on the multiple realization of JAR in one direction or the other.

Additional reading

Aizawa, K. forthcoming. Multiple Realization, Autonomy, and Integration. In D. Kaplan (ed.), *Integrating Mind and Brain Science: Mechanistic Perspectives and Beyond.* Oxford: Oxford University Press.

Aizawa, K. and C. Gillett. 2011. The Autonomy of Psychology in the Age of Neuroscience. In P. McKay Illari, F. Russo, and J. Williamson (eds), *Causality in the Sciences.* Oxford: Oxford University Press: 202–23.

Craver, C. 2004. Dissociable Realization and Kind Splitting. *Philosophy of Science* 71: 960–71.

Keeley, B. 2000. Shocking Lessons from Electric Fish: The Theory and Practice of Multiple Realization. *Philosophy of Science* 67 (3): 444–65.

Chapter 7

1. Consider the statement "Bigfoot just ducked behind the tree." What is wrong with accepting this as evidence in favor of the hypothesis that Bigfoot exists? How does your answer to this question bear on using plasticity as evidence in favor of the hypothesis of multiple realization?

2. Why might the phenomenon of convergent evolution lead one to expect multiple realizability? Are Polger and Shapiro right that complex biological systems are highly constrained?

3. Why is the distinction between *causes* of behavior and *psychological causes* of behavior important for evaluating the significance of convergent evolution for multiple realizability?

4. What does the possibility of future artificial intelligences tell us about the prospects for multiple realizability?

Additional reading

Block, N. and J. Fodor. 1972. What Psychological States Are Not. *Philosophical Review* 81: 159–81.

Figdor, C. 2010. Neuroscience and the Multiple Realization of Cognitive Functions. *Philosophy of Science* 77 (3): 419–56.

Shapiro, L. 2004. *The Mind Incarnate*. Cambridge, MA: MIT Press: Chs. 3–4.

Sober, E. 2008. *Evidence and Evolution: The Logic behind the Science*. Cambridge: Cambridge University Press.

Weiskopf, D. 2011. The Functional Unity of Special Science Kinds. *British Journal for the Philosophy of Science* 62 (2): 233–58.

Chapter 8

1. Trace two different versions of the line of reasoning that might take one from belief that cognition is computational to the conclusion that cognitive processes are multiply realizable.

2. Granting that some cognitive systems might be given a computational description, what follows about the appropriateness of describing the processes within the system algorithmically? Which considerations might favor an algorithmic description, and which not?

3. Explain the distinction between an algorithmic description that is "ontologically committing" and one that is not. What has this to do with the possibility of multiple realization?

4. When Fodor claims that computationalism is the only game in town, what sort of computationalism does he have in mind? Is it the only game in town?

Additional reading

Bechtel, W. and R. McCauley. 1999. Heuristic Identity Theory (or Back to the Future): The Mind-Body Problem against the Background of Research Strategies in Cognitive Neuroscience. *Proceedings of the 21st Annual Meeting of the Cognitive Science Society*: 67–72.

Bechtel, W. and J. Mundale. 1999. Multiple Realizability Revisited: Linking Cognitive and Neural States. *Philosophy of Science* 66: 175–207.

Chemero, A. 2009. *Radical Embodied Cognitive Science*. Cambridge, MA: MIT Press.

Fodor, J. 1975. *The Language of Thought*. New York: Thomas Crowell and Co.

Haugeland, J. 1981. Semantic Engines: An Introduction to Mind Design. In J. Haugeland (ed.), *Mind Design*, 1st ed. Cambridge, MA: MIT Press: 1–34.

Piccinini, G. and A. Scarantino. 2011. Information Processing, Computation, and Cognition. *Journal of Biological Physics* 37 (1): 1–38.

Chapter 9

1. What are "unification strategies" for challenging claims of multiple realization and what do they entail about the elimination of psychological states?

2. What do abstraction and idealization involve, and how might they result in the false impression of multiple realization? Do abstraction and idealization force scientists to be eliminativists?

3. Why might someone think that the use of heuristics in the cognitive sciences leads to accidental eliminativism?

4. How might kind splitting lead to eliminativism? Must it?

Additional reading

Bechtel, W. and R. McCauley. 1999. Heuristic Identity Theory (or Back to the Future): The Mind-Body Problem against the Background of Research Strategies in Cognitive Neuroscience. *Proceedings of the 21st Annual Meeting of the Cognitive Science Society*: 67–72.

Klein, C. 2008. An Ideal Solution to Disputes about Multiply Realized Kinds. *Philosophical Studies* 140 (2): 161–77.

Potochnik, A. Forthcoming. *Idealization and the Aims of Science*. Chicago, IL: Chicago University Press.

Weisberg, M. 2007. Three Kinds of Idealization. *Journal of Philosophy* 104 (12): 639–59.

Chapter 10

1. What is Descartes' Revenge and why might some, e.g. Kim, think that an identity theory avoids the problem?

2. How does the problem of causal exclusion differ from the problem of explanatory exclusion?

3. Explain the idea behind Essential Autonomy. What consequences do realizationist theorists draw from this principle about the relationship between psychology and neuroscience?

4. What is meant by the Actual Autonomy of an explanation, and how does the idea of Actual Autonomy differ from Essential Autonomy with respect to questions about the legitimacy of explanations?

5. Why is Kim skeptical of the value of identity claims in explanations? What value might such claims have?

Additional reading

Antony, L. and J. Levine. 1997. Reduction with Autonomy. In Tomberlin 1997.

Bennett, K. 2008. Exclusion Again. In J. Hohwy and J. Kallestrup (eds), *Being Reduced*. Oxford: Oxford University Press.

Fodor, J. 1997. Special Sciences: Still Autonomous after All These Years. In Tomberlin 1997.

Kim, J. 1998. *Mind in a Physical World*. Cambridge, MA: MIT Press.

Kim, J. 2005. Physicalism, or Something Near Enough. Princeton, NJ: Princeton University Press.

References

Aizawa, K. 2007. The Biochemistry of Memory Consolidation: A Model System for the Philosophy of Mind. *Synthese* 155 (1): 65–98.

Aizawa, K. 2008. Neuroscience and Multiple Realization: A Reply to Bechtel and Mundale. *Synthese* 167: 495–510.

Aizawa, K. 2010. Computation in Cognitive Science: It Is Not All about Turing-Equivalent Computation. *Studies in History and Philosophy of Science Part A* 41 (3): 227–36.

Aizawa, K. 2013. Multiple Realizability by Compensatory Differences. *European Journal for Philosophy of Science* 3 (1): 69–86.

Aizawa, K. Forthcoming. Multiple Realization, Autonomy, and Integration. In D. Kaplan (ed.), *Integrating Mind and Brain Science: Mechanistic Perspectives and Beyond*. Oxford: Oxford University Press.

Aizawa, K. and C. Gillett. 2009a. The (Multiple) Realization of Psychological and Other Properties in the Sciences. *Mind and Language* 24: 181–208.

Aizawa, K. and C. Gillett. 2009b. Levels, Individual Variation, and Massive Multiple Realization in Neurobiology. In J. Bickle (ed.), *Oxford Handbook of Philosophy and Neuroscience*. New York: Oxford University Press: 529–81.

Aizawa, K. and C. Gillett. 2011. The Autonomy of Psychology in the Age of Neuroscience. In P. McKay Illari, F. Russo, and J. Williamson (eds), *Causality in the Sciences*. Oxford: Oxford University Press: 202–23.

Alupay, J., S. Hadjisolomou, and R. Crook. 2014. Arm Injury Produces Long-Term Behavioral and Neural Hypersensitivity in Octopus. *Neuroscience Letters* 558: 137–42.

Anderson, M. 2010. Neural Reuse: A Fundamental Organizational Principle of the Brain. *Behavioral and Brain Sciences* 33: 245–313.

Anderson, M. 2014. *After Phrenology: Neural Reuse and the Interactive Brain*. Cambridge, MA: MIT Press.

Antony, L. 1995. Law and Order in Psychology. *Philosophical Perspectives* 9: 429–46.

Antony, L. 1999. Multiple Realizability, Projectibility, and the Reality of Mental Properties. *Philosophical Topics* 26 (1/2): 1–24.

Antony, L. 2003. Who's Afraid of Disjunctive Properties? *Philosophical Issues* 13 (1): 1–21.

Antony, L. 2008. Multiple Realization: Keeping It Real. In J. Hohwy and J. Kallestrup (eds), *Being Reduced: New Essays on Reduction, Explanation, and Causation*. Oxford: Oxford University Press.

Antony, L. 2010. Realization Theory and the Philosophy of Mind: Comments on Sydney Shoemaker's Physical Realization. *Philosophical Studies* 148 (1): 89–99.

Antony, L. and J. Levine. 1997. Reduction with Autonomy. In Tomberlin 1997.

Avarguès-Weber, A., G. Portelli, J. Benard, A. Dyer, and M. Giurfa. 2010. Configural Processing Enables Discrimination and Categorization of Face-Like Stimuli in Honeybees. *Journal of Experimental Biology* 213 (4): 593–601.

Bach-y-Rita, P. and S. Kercel. 2003. Sensory Substitution and the Human-Machine Interface. *Trends in Cognitive Sciences* 7 (12): 541–6.

Balari, S. and G. Lorenzo. 2015. Ahistorical Homology and Multiple Realizability. *Philosophical Psychology* 28 (6): 881–902.

Barrett, D. 2013. Multiple Realizability, Identity Theory, and the Gradual Reorganization Principle. *British Journal for the Philosophy of Science* 64 (2): 325–46.

Barrett, D. 2014. Functional Analysis and Mechanistic Explanation. *Synthese* 191 (12): 2695–714.

Baumgartner, M. 2009. Interventionist Causal Exclusion and Non-Reductive Physicalism. *International Studies in the Philosophy of Science* 23 (2): 161–78.

Baumgartner, M. 2010. Interventionism and Epiphenomenalism. *Canadian Journal of Philosophy* 40: 359–83.

Baumgartner, M. 2013. Rendering Interventionism and Non-Reductive Physicalism Compatible. *Dialectica* 67: 1–27.

Bechtel, W. 2008. *Mental Mechanisms: Philosophical Perspectives on Cognitive Neuroscience*. London: Routledge.

Bechtel, W. and A. Abrahamsen. 1990. *Connectionism and the Mind: An Introduction to Parallel Processing in Networks*. Cambridge, MA: Blackwell.

Bechtel, W. and R. McCauley. 1999. Heuristic Identity Theory (or Back to the Future): The Mind-Body Problem against the Background of Research Strategies in Cognitive Neuroscience. *Proceedings of the 21st Annual Meeting of the Cognitive Science Society*: 67–72.

Bechtel, W. and J. Mundale. 1999. Multiple Realizability Revisited: Linking Cognitive and Neural States. *Philosophy of Science* 66: 175–207.

Bechtel, W. and R. Richardson. 1993. *Discovering Complexity: Decomposition and Localization as Strategies in Scientific Research*. Princeton, NJ: Princeton University Press.

Bechtel, W. and R. Richardson. 2010. *Discovering Complexity: Decomposition and Localization as Strategies in Scientific Research*. 2nd Ed. Cambridge, MA: MIT Press.

Beer, R. 1990. *Intelligence as Adaptive Behavior: An Experiment in Computational Neuroethology*. Boston, MA: Academic Press.

Bennett, K. 2003. Why the Exclusion Problem Seems Intractable and How, Just Maybe, to Tract It. *Noûs* 37 (3): 471–97.

Bennett, K. 2008. Exclusion Again. In J. Hohwy and J. Kallestrup (eds), *Being Reduced*. Oxford: Oxford University Press.

Bennett, K. 2011. Construction Area (No Hard Hat Required). *Philosophical Studies* 154 (1): 79–104.

Bennett, K. Forthcoming. *Making Things Up*. Oxford: Oxford University Press.

Bickle, J. 1998. *Psychoneural Reduction*. Cambridge, MA: MIT Press.

Bickle, J. 2003. *Philosophy and Neuroscience: A Ruthlessly Reductive Account*. Boston, MA: Kluwer Academic Publishers.

Bickle, J. 2006. Reducing Mind to Molecular Pathways: Explicating the Reductionism Implicit in Current Cellular and Molecular Neuroscience. *Synthese* 151: 411–34.

Block, N. 1978. Troubles with Functionalism. In C. W. Savage (ed.), *Minnesota Studies in the Philosophy of Science*, Vol. IX. Minneapolis, MN: University of Minnesota Press. Reprinted in Block 1980a.

Block, N. (ed.) 1980a. *Readings in Philosophy of Psychology*, Volume One. Cambridge, MA: Harvard University Press.

Block, N. (ed.) 1980b. *Readings in Philosophy of Psychology*, Volume Two. Cambridge, MA: Harvard University Press.

Block, N. 1997. Anti-Reductionism Slaps Back. In Tomberlin 1997.

Block, N. 2003. Do Causal Powers Drain Away? *Philosophy and Phenomenological Research* 67: 133–50.

Block, N. and J. Fodor. 1972. What Psychological States Are Not. *Philosophical Review* 81: 159–81.

Block, N. and R. Stalnaker. 1999. Conceptual Analysis, Dualism, and the Explanatory Gap. *Philosophical Review* 108 (1): 1–46.

Bontly, T. 2005. Exclusion, Overdetermination, and the Nature of Causation. *Journal of Philosophical Research* 30: 261–82.

Buonomano, D. and M. Merzenich. 1998. Cortical Plasticity: From Synapses to Maps. *Annual Review of Neuroscience* 21: 149–86.

Calvo Garzón, P. and F. Keijzer. 2009. Cognition in Plants. In F. Baluška (ed.), *Plant-Environment Interactions*. Berlin Heidelberg: Springer-Verlag: 247–66.

Cao, R. 2012. A Teleosemantic Approach to Information in the Brain. *Biology and Philosophy* 27 (1): 49–71.

Cartwright, N. 1983. *How the Laws of Physics Lie*. Oxford: Oxford University Press.

Chalmers, D. 1996. Does a rock implement every finite-state automaton? *Synthese* 108: 309–33.

Chatham, C. and D. Badre. Forthcoming. How to Test Cognitive Theory with fMRI. In E. Schumacher and D. Spieler (eds), *New Methods in Cognitive Psychology*. Hove: Psychology Press.

Chemero, A. 2009. *Radical Embodied Cognitive Science*. Cambridge, MA: MIT Press.

Chirimuuta, M. 2014. Minimal Models and Canonical Neural Computations: The Distinctness of Computational Explanation in Neuroscience. *Synthese* 191 (2): 127–53.

Chomsky, N. 1957. *Syntactic Structures*. The Hague: Mouton.

Chomsky, N. 1959. A Review of B. F. Skinner's Verbal Behavior. *Language* 35 (1): 26–58.

Churchland, P. M. 1981. Eliminative Materialism and the Propositional Attitudes. *Journal of Philosophy* 78 (2): 67–90.

Churchland, P. M. 1982. Is "Thinker" a Natural Kind? *Dialogue* 21 (2): 223–38.

Churchland, P. S. 1983. Consciousness: The Transmutation of a Concept. *Pacific Philosophical Quarterly* 64: 80–93.

Churchland, P. S. 1986. *Neurophilosophy: Toward a Unified Science of Mind-Brain*. Cambridge, MA: MIT Press.

Churchland, P. S., V. Ramachandran, and T. Sejnowski. 1994. A Critique of Pure Vision. In C. Koch and J. Davis (eds), *Large-Scale Neuronal Theories of the Brain*. Cambridge, MA: MIT Press: 23–60.

Clapp, L. 2001. Disjunctive Properties: Multiple Realizations. *Journal of Philosophy* 3: 111–36.

Clark, A. 1993. *Associative Engines*. Cambridge, MA: MIT Press.

Couch, M. 2004. Discussion: A Defense of Bechtel and Mundale. *Philosophy of Science* 71 (2): 198–204.

Couch, M. 2005. Functional Properties and Convergence in Biology. *Philosophy of Science* 72 (5): 1041–51.

Couch, M. 2009. Multiple Realization in Comparative Perspective. *Biology and Philosophy* 24 (4): 505–19.

Craver, C. 2001. Role Functions, Mechanisms, and Hierarchy. *Philosophy of Science* 68: 53–74.

Craver, C. 2002. Interlevel Experiments and Multilevel Mechanisms in the Neuroscience of Memory. *Philosophy of Science (Supplement)* 69 (3): S83–97.

Craver, C. 2004. Dissociable Realization and Kind Splitting. *Philosophy of Science* 71: 960–71.

Craver, C. 2007. *Explaining the Brain: Mechanisms and the Mosaic Unity of Neuroscience*. Oxford: Oxford University Press.

Craver, C. and W. Bechtel. 2007. Top-Down Causation without Top-Down Causes. *Biology and Philosophy* 20: 715–34.

Craver, C. and L. Darden. 2001. Discovering Mechanisms in Neuroscience: The Case of Spatial Memory. In P. Machamer, R. Gush, and P. McLaughlin (eds), *Theory and Method in Neuroscience*. Pittsburgh, PN: University of Pittsburgh Press.

Craver, C. and L. Darden. 2013. *In Search of Mechanisms: Discoveries across the Life Sciences*. Chicago, IL: University of Chicago Press.

Cummins, R. 2000. "How Does It Work" versus "What Are the Laws?": Two Conceptions of Psychological Explanation. In F. Keil and R. Wilson (eds), *Explanation and Cognition*. Cambridge, MA: MIT Press: 117–45.

Darwin, C. 1859/1964. *On the Origin of Species: A Facsimile of the First Edition*. Cambridge, MA: Harvard University Press.

David, M. 1997. Kim's Functionalism. In Tomberlin 1997.

Davidson, D. 1970. Mental Events. In L. Foster and J. W. Swanson (eds), *Experience and Theory*. Amherst, MA: University of Massachusetts Press.

Dennett, D. 1984. *Elbow Room*. Oxford: Oxford University Press.

Dennett, D. 1987. True Believers. In D. Dennett, *The Intentional Stance*. Cambridge, MA: MIT Press.

Dugas-Ford, J., J. Rowell, and C. Ragsdale. 2012. Cell-Type Homologies and the Origins of the Neocortex. *Proceedings of the National Academy of Sciences* 109: 16974–9.

Duhem, P. 1914/1954. *The Aim and Structure of Physical Theory*. Princeton, NJ: Princeton University Press.

Dyer, A. 2012. The Mysterious Cognitive Abilities of Bees: Why Models of Visual Processing Need to Consider Experience and Individual Differences in Animal Performance. *Journal of Experimental Biology* 215: 387–95.

Dyer, A., C. Neumeyer, and L. Chittka. 2005. Honeybee (*Apis mellifera*) Vision Can Discriminate between and Recognise Images of Human Faces. *Journal of Experimental Biology* 208: 4709–14.

Edelman, G. and J. Gally. 2001. Degeneracy and Complexity in Biological Systems. *Proceedings of the National Academy of Sciences* 98 (24): 13763–8.

Elgin, C. 2004. True Enough. *Philosophical Issues* 14: 113–31.

Eliasmith, C. 2010. How We Ought to Describe Computation in the Brain. *Studies in the History and Philosophy of Science* 41: 313–20.

Endicott, R. 2010. Realization, Reductios, and Category Inclusion. *Journal of Philosophy* 107 (4): 213–19.

Endicott, R. 2012. Resolving Arguments by Different Conceptual Traditions of Realization. *Philosophical Studies* 159 (1): 41–59.

Eronen, M. 2012. Pluralistic Physicalism and the Causal Exclusion Argument. *European Journal for the Philosophy of Science* 2: 219–32.

Feigl, H. 1958. The "Mental" and the "Physical." *Minnesota Studies in the Philosophy of Science* 2: 370–497.

Fernald, R. 2000. Evolution of Eyes. *Current Opinions in Neurobiology* 10: 444–50.

Figdor, C. 2010. Neuroscience and the Multiple Realization of Cognitive Functions. *Philosophy of Science* 77 (3): 419–56.

Fine, K. 1995. Ontological Dependence. *Proceedings of the Aristotelian Society* 95: 269–90.

Fine, K. 2001. The Question of Realism. *Philosophers' Imprint* 1 (1): 1–30.

Fodor, J. 1968. *Psychological Explanation*. New York: Random House.

Fodor, J. 1974. Special Sciences, or the Disunity of Science as a Working Hypothesis. *Synthese* 28: 97–115.

Fodor, J. 1975. *The Language of Thought*. New York: Thomas Crowell and Co.

Fodor, J. 1981. The Mind-Body Problem. *Scientific American* 244: 114–25.

Fodor, J. 1983. *The Modularity of Mind*. Cambridge, MA: MIT Press.

Fodor, J. 1985. Fodor's Guide to Mental Representation: The Intelligent Auntie's Vade-Mecum. *Mind*: 55–97.

Fodor, J. 1989. Making Mind Matter More. *Philosophical Topics* LXV (1): 59–79.

Fodor, J. 1997. Special Sciences: Still Autonomous after All These Years. In Tomberlin 1997.

Fodor, J. 1999. Diary: Why the Brain? *London Review of Books* 21 (19): 68–9.

Francescotti, R. 1997. What Multiple Realizability Does Not Show. *Journal of Mind and Behavior* 18 (1): 13–28.

Francescotti, R. 2014. *Physicalism and the Mind*. Dordrecht: Springer.

Fumerton, R. 2007. Render unto Philosophy that Which Is Philosophy's. *Midwest Studies in Philosophy* XXXI: 56–67.

Funkhouser, E. 2007. A Liberal Conception of Multiple Realizability. *Philosophical Studies* 132 (3): 467–94.

Funkhouser, E. 2014. *The Logical Structure of Kinds*. Oxford: Oxford University Press.

Garson, J. 2003. The Introduction of Information into Neurobiology. *Philosophy of Science* 70 (5): 926–36.

Gibson, J. J. 1950. *The Perception of the Visual World*. Boston, MA: Houghton Mifflin.

Gibson, J. J. 1966. *The Senses Considered as Perceptual Systems*. Prospect Heights, NY: Waveland Press.

Gibson, J. J. 1979. *The Ecological Approach to Visual Perception*. Boston, MA: Houghton-Mifflin.

Gillett, C. 2002. The Dimensions of Realization: A Critique of the Standard View. *Analysis* 64 (4): 316–23.

Gillett, C. 2003. The Metaphysics of Realization, Multiple Realizability and the Special Sciences. *Journal of Philosophy* 100 (11): 91–603.

Gillett, C. 2007. Understanding the New Reductionism: The Metaphyics of Science and Compositional Reduction. *Journal of Philosophy* 104 (4): 193–216.

Glennan, S. 1996. Mechanisms and the Nature of Causation. *Erkenntnis* 44: 49–71.

Glennan, S. 2002. Rethinking Mechanistic Explanation. *Philosophy of Science* 69: S342–53.

Glennan, S. 2005. Modeling Mechanisms. *Studies in History and Philosophy of Biological and Biomedical Sciences* 36: 443–64.

Godfrey-Smith, P. 2005. Folk Psychology as a Model. *Philosophers Imprint* 5 (6): 1–16.

Godfrey-Smith, P. 2006. The Strategy of Model-Based Science. *Biology and Philosophy* 21: 725–40.

Gozzano, S. 2010. Multiple Realizability and Mind-Body Identity. In M. Suarez, M. Dorato, and M. Redei (eds), *EPSA: Epistemology and Methodology of Science*. New York: Springer: 119–27.

Guttenplan, S. (ed.). 1994. *A Companion to the Philosophy of Mind*. Oxford: Blackwell.

Hardin, C. 1988. *Color for Philosophers*. Indianapolis, IN: Hackett.

Harris, W. 1997. Pax-6: Where to Be Conserved Is Not Conservative. *Proceedings of the National Academy of Sciences* 94 (6): 2098–100.

Hatfield, G. 1991. Representation and Rule-Instantiation in Connectionist Systems. In T. Horgan and J. Tienson (eds), *Connectionism and the Philosophy of Mind*. Boston, MA: Kluwer: 90–112.

Haug, M. 2010. Realization, Determination, and Mechanisms. *Philosophical Studies* 150 (3): 313–30.

Haug, M. 2011a. Abstraction and Explanatory Relevance; or, Why Do the Special Sciences Exist? *Philosophy of Science* 78 (5): 1143–55.

Haug, M. 2011b. Natural Properties and the Special Sciences. *Monist* 94 (2): 244–66.

Haugeland, J. 1981. Semantic Engines: An Introduction to Mind Design. In J. Haugeland (ed.), *Mind Design*, 1st. ed. Cambridge, MA: MIT Press: 1–34.

Haugeland, J. 1985. *Artificial Intelligence: The Very Idea*. Cambridge, MA: MIT Press.

Heil, J. 1999. Multiple Realizability. *American Philosophical Quarterly* 36 (3): 189–208.

Heiligenberg, W. 1991. The Jamming Avoidance Response of the Electric Fish, Eigenmannia: Computational Rules and Neuronal Implementation. *Seminars in the Neurosciences* 3: 3–18.

Hempel, C. and P. Oppenheim. 1948. Studies in the Logic of Explanation. *Philosophy of Science* 15 (2): 135–75.

Hill, C. 1991. *Sensations: A Defense of Type Materialism*. Cambridge: Cambridge University Press.

Hochner, B., T. Shomrat, and G. Fiorito. 2006. The Octopus: A Model for a Comparative Analysis of the Evolution of Learning and Memory Mechanisms. *Biological Bulletin* 210: 308–17.

Horgan, T. 1993a. From Supervenience to Superdupervenience: Meeting the Demands of a Material World. *Mind* 102 (408): 555–86.

Horgan, T. 1993b. Nonreductive Materialism and the Explanatory Autonomy of Psychology. In R. Wagner and S. J. Warner (eds), *Naturalism: A Critical Appraisal*. Notre Dame, IN: University of Notre Dame Press.

Jackson, F. 1998. *From Metaphysics to Ethics: A Defense of Conceptual Analysis.* Oxford: Oxford University Press.

Jackson, F. and P. Pettit. 1995. Moral Functionalism and Moral Motivation. *Philosophical Quarterly* 45 (178): 20–40.

Jaworksi, W. 2002. Multiple-Realizability, Explanation, and the Disjunctive Move. *Philosophical Studies* 108 (3): 298–308.

Kaas, J. 1991. Plasticity of Sensory and Motor Maps in Adult Mammals. *Annual Review of Neuroscience* 14: 137–67.

Kallestrup, J. 2006. The Causal Exclusion Argument. *Philosophical Studies* 131 (2): 459–85.

Kaplan, D. and C. Craver. 2011. The Explanatory Force of Dynamical and Mathematical Models in Neuroscience: A Mechanistic Perspective. *Philosophy of Science* 78: 601–27.

Karten, H. 2013. Neocortical Evolution: Neuronal Circuits Arise Independently of Lamination. *Current Biology* 23 (1): R12–15.

Kawasaki, M. 1993. Independently Evolved Jamming Avoidance Responses Employ Identical Computational Algorithms: A Behavioral Study of the African Electric Fish, *Gymnarchus niloticus. Journal of Comparative Physiology A* 173: 9–22.

Kawasaki, M. 2009. Evolution of Time-Coding Systems in Weakly Electric Fishes. *Zoological Science* 26 (9): 587–9.

Keaton, D. and T. Polger. 2014. Exclusion, Still Not Tracted. *Philosophical Studies* 171 (1): 135–48.

Keeley, B. 2000. Shocking Lessons from Electric Fish: The Theory and Practice of Multiple Realization. *Philosophy of Science* 67 (3): 444–65.

Keeley, B. 2002. Making Sense of the Senses: Individuating Modalities in Humans and Other Animals. *Journal of Philosophy* 99 (1): 5–28.

Keijzer, F. 2013. The Sphex Story: How the Cognitive Sciences Kept Repeating an Old and Questionable Anecdote. *Philosophical Psychology* 26 (4): 502–19.

Kelso, J. A. 1995. *Dynamic Patterns: The Self-Organization of Brain and Behavior.* Cambridge, MA: MIT Press.

Kim, J. 1972. Phenomenal Properties, Psychophysical Laws, and Identity Theory. *Monist* 56 (2): 177–92. Excerpted in Block 1980a under the title, "Physicalism and the Multiple Realizability of Mental States."

Kim, J. 1989. The Myth of Nonreductive Materialism. *Proceedings and Addresses of the American Philosophical Association* 63 (3): 31–47.

Kim, J. 1997. The Mind-Body Problem: Taking Stock after Forty Years. In Tomberlin 1997.

Kim, J. 1998. *Mind in a Physical World.* Cambridge, MA: MIT Press.

Kim, J. 1999. Hempel, Explanation, and Metaphysics. *Philosophical Studies* 94: 1–20.

Kim, J. 2005. Physicalism, or Something Near Enough. Princeton, NJ: Princeton University Press.

Kim, J. 2012. The Very Idea of Token Physicalism. In S. Gozzano and C. Hill (eds), *New Perspectives on Type Identity: The Mental and the Physical*. Cambridge: Cambridge University Press: 167–85.

Kim, S. 2002. Testing Multiple Realizability: A Discussion of Bechtel and Mundale. *Philosophy of Science* 69 (4): 606–10.

Kim, S. 2009. Multiple Realizations, Diverse Implementations and Antireductionism. *Theoria* 75 (3): 232–44.

Kim, S. 2011. Multiple Realization and Evidence. *Philosophical Psychology* 24 (6): 739–49.

Kitcher, P. 1980. How to Reduce a Functional Psychology. *Philosophy of Science* 47 (1): 134–40.

Kitcher, P. 1984. 1953 and All That: A Tale of Two Sciences. *Philosophical Review* 93 (3): 335–73.

Klein, C. 2008. An Ideal Solution to Disputes about Multiply Realized Kinds. *Philosophical Studies* 140 (2): 161–77.

Klein, C. 2012. Cognitive Ontology and Region versus Network-Oriented Analyses. *Philosophy of Science* 79 (5): 952–60.

Klein, C. 2013. Multiple Realizability and the Semantic View of Theories. *Philosophical Studies* 163 (3): 683–95.

Klein, C. 2014. Psychological Explanation, Ontological Commitment, and the Semantic View of Theories. In M. Sprevak and J. Kallestrup (eds), *New Waves in Philosophy of Mind*. New York: Palgrave Macmillan: 208–25.

Klinge, C., F. Eippert, B. Roder, and C. Buchel. 2010. Corticocortical Connections Mediate Primary Visual Cortex Responses to Auditory Stimulation in the Blind. *Journal of Neuroscience* 30: 12798–805.

Kobes, B. 1991. On a Model for Psycho-Neural Coevolution. *Behavior and Philosophy* 19: 1–17.

Kosslyn, S. and G. Hatfield. 1984. Representation without Symbol Systems. *Social Research* 51: 1019–45.

Lamb, M. 2015. Characteristics of Non-Reductive Explanations in Complex Dynamical Systems Research. PhD dissertation, University of Cincinnati.

Land, M. and R. Fernald. 1992. The Evolution of Eyes. *Annual Review of Neuroscience* 15: 1–29.

Lange, M. Forthcoming. Two Kinds of Identity Explanations. Presented at the 2014 meeting of the Philosophy of Science Association.

Lashley, K. 1929. *Brain Mechanisms and Intelligence*. Chicago, IL: University of Chicago Press.

Lewis, D. 1969. Review of Art, Mind, and Religion. *Journal of Philosophy* 66: 23–35. Excerpted in Block 1980a as "Review of Putnam."

Lewis, D. 1970. How to Define Theoretical Terms. *Journal of Philosophy* 68: 203–11.

Lewis, D. 1972. Psychophysical and Theoretical Identifications. *Australasian Journal of Philosophy* 50: 249–58.

Lewis, D. 1980. Mad Pain and Martian Pain. In Block 1980a.

Lewis, D. 1994. Lewis, David: Reduction of Mind. In Guttenplan 1994.

List, C. and P. Menzies. 2009. Nonreductive Physicalism and the Limits of the Exclusion Principle. *Journal of Philosophy* 106: 475–502.

Lycan, W. 1987. *Consciousness*. Cambridge, MA: MIT Press.

Lycan, W. 1990. *Mind and Cognition: A Reader*. Cambridge, MA: Blackwell.

Lynch, M. 2000. Alethic Pluralism and the Functionalist Theory of Truth. *Acta Analytica* 15: 195–214.

Lynch, M. 2004. *True to Life*. Cambridge, MA: MIT Press.

Lyre, H. 2009. The "Multirealization" of Multiple Realizability. In A. Hieke and H. Leitgeb (eds), *Reduction-Abstraction-Analysis*. Berlin: Ontos: 79–94.

Machamer, P., L. Darden, and C. Craver. 2000. Thinking about Mechanisms. *Philosophy of Science* 67: 1–25.

MacQuorcodale, K. 1970. A Reply to Chomsky's Review of Skinner's Verbal Behavior. *Journal of the Experimental Analysis of Behavior* 13: 83–99.

Marr, D. 1982. *Vision: A Computational Investigation into the Human Representation and Processing of Visual Information*. San Francisco, CA: W. H. Freeman.

McCaffrey, J. 2015. The Brain's Heterogeneous Functional Landscape. *Philosophy of Science* 82 (5): 1010–22.

McCauley, R. 2012. About Face: Philosophical Naturalism, Heuristic Identity Theory, and Recent Findings about Prosopagnosia. In S. Gozzano and C. Hill (eds), *New Perspectives on Type Identity: The Mental and the Physical*. Cambridge: Cambridge University Press: 186–206.

McCauley, R. and W. Bechtel. 2001. Explanatory Pluralism and Heuristic Identity Theory. *Theory Psychology* 11 (6): 736–60.

McClelland, J., D. Rumelhart, and the PDP Research Group. 1986. *Parallel Distributed Processing: Explorations in the Microstructure of Cognition*, Volume II. Cambridge, MA: MIT Press.

McLaughlin, B. 2001. In Defense of New Wave Materialism: A Response to Horgan and Tienson. In C. Gillett and B. Loewer (eds), *Physicalism and Its Discontents*. Cambridge: Cambridge University Press.

Melnyk, A. 2003. *A Physicalist Manifesto: Thoroughly Modern Materialism*. Cambridge: Cambridge University Press.

Melzack, R. and W. Torgerson. 1971. On the Language of Pain. *Anesthesiology* 34: 50–9.

Menzel, R. and M. Giurfa. 2006. Dimensions of Cognition in an Insect, the Honeybee. *Behavioral and Cognitive Neuroscience Reviews* 5 (1): 24–40.

Miłkowski, M. 2013a. *Explaining the Computational Mind*. Cambridge, MA: MIT Press.

Miłkowski, M. 2013b. A Mechanistic Account of Computational Explanation in Cognitive Science. In M. Knauff, M. Pauen, N. Sebanz, and I. Wachsmuth (eds), *Cooperative Minds: Social Interaction and Group Dynamics. Proceedings of the 35th Annual Meeting of the Cognitive Science Society*. Austin, TX: Cognitive Science Society: 3050–5.

Miłkowski, M. 2015. Evaluating Artificial Models of Cognition. *Studies in Logic, Grammar, and Rhetoric* 40 (53): 43–62.

Millikan, R. 1984. *Language, Thought, and Other Biological Categories*. Cambridge, MA: MIT Press.

Millikan, R. 1989. In Defense of Proper Functions. *Philosophy of Science* 56: 288–302. Reprinted in Millikan 1993: 13–29.

Millikan, R. 1993. *White Queen Psychology and Other Essays for Alice*. Cambridge, MA: MIT Press.

Millikan, R. 1999. Historical Kinds and the "Special Sciences." *Philosophical Studies* 95: 45–65.

Morris, K. 2010. Guidelines for Theorizing about Realization. *Southern Journal of Philosophy* 48 (4): 393–416.

Morris, K. 2011a. Subset Realization, Parthood, and Causal Overdetermination. *Pacific Philosophical Quarterly* 92 (3): 363–79.

Morris, K. 2011b. Theoretical Identities as Explanantia and Explananda. *American Philosophical Quarterly* 48 (4): 373–85.

Morris, K. 2014. Causal Closure, Causal Exclusion, and Supervenience Physicalism. *Pacific Philosophical Quarterly* 95 (1): 72–86.

Morris, K. 2015. Against Disanalogy-Style Responses to the Exclusion Problem. *Philosophia* 43 (2): 435–53.

Mundale, J. 2002. Concepts of Localization: Balkanization in the Brain. *Brain and Mind* 3: 1–18.

Newell, A. and H. Simon. 1972. *Human Problem Solving*. Englewood Cliffs, NJ: Prentice-Hall.

Ney, A. 2007. Can an Appeal to Constitution Solve the Exclusion Problem? *Pacific Philosophical Quarterly* 88 (4): 486–506.

Ney, A. 2009. Physical Causation and Difference-Making. *British Journal for the Philosophy of Science* 60 (4): 737–64.

Ney, A. 2012. The Causal Contribution of Mental Events. In S. Gozzano and C. Hill (eds), *New Perspectives on Type Identity: The Mental and the Physical*. Cambridge: Cambridge University Press: 230–50.

Nilsson, D. and S. Pelger. 1994. A Pessimistic Estimate of the Time Required for an Eye to Evolve. *Proceedings of the Royal Society* 256 (1345).

Noë, A. 2005. *Action in Perception*. Cambridge, MA: MIT Press.

Noppeney, U., K. Friston, and C. Price. 2004. Degenerate Neuronal Systems Sustaining Cognitive Functions. *Journal of Anatomy* 205: 433–42.

Oppenheim, P. and H. Putnam. 1958. The Unity of Science as a Working Hypothesis. In H. Feigl et al. (eds), *Minnesota Studies in the Philosophy of Science*, Volume 2. Minneapolis, MN: Minnesota University Press.

Orlandi, N. 2014. *The Innocent Eye: Why Vision Is Not a Cognitive Process*. Oxford: Oxford University Press.

Palmer, D. C. 2006. On Chomsky's Appraisal of Skinner's *Verbal Behavior*: A Half Century of Misunderstanding. *Behavior Analyst* 29 (2): 253–67.

Papineau, D. 2001. The Rise of Physicalism. In C. Gillett and B. M. Loewer (eds), *Physicalism and Its Discontents*. Cambridge: Cambridge University Press.

Pascalis, O., D. Kelly, and R. Caldara. 2006. What Can Bees Really Tell Us about the Face Processing System in Humans? *Journal of Experimental Biology* 209: 3266.

Pauen, M. 2002. Is Type Identity Incompatible with Multiple Realization? *Grazer Philosophische Studien* 65 (1): 37–49.

Pearl, J. 2000. *Causality*. Cambridge: Cambridge University Press.

Piatigorsky, J. 2008. A Genetic Perspective on Eye Evolution: Gene Sharing, Convergence and Parallelism. *Evolution: Education and Outreach* 1: 403–14.

Piccinini, G. 2007. Computing Mechanisms. *Philosophy of Science* 74 (4): 501–26.

Piccinini, G. 2015. *Physical Computation*. Oxford: Oxford University Press.

Piccinini, G. and S. Bahar. 2013. Neural Computation and the Computational Theory of Cognition. *Cognitive Science* 37 (3): 453–88.

Piccinini, G. and C. Craver. 2011. Integrating Psychology and Neuroscience: Functional Analyses as Mechanism Sketches. *Synthese* 183: 283–311.

Piccinini, G. and A. Scarantino. 2010. Computation vs. Information Processing: Why Their Difference Matters to Cognitive Science. *Studies in History and Philosophy of Science* 41: 237–46.

Piccinini, G. and A. Scarantino. 2011. Information Processing, Computation, and Cognition. *Journal of Biological Physics* 37 (1): 1–38.

Piccinini, G. and O. Shagrir. 2014. Foundations of Computational Neuroscience. *Current Opinion in Neurobiology* 25: 25–30.

Place, U. T. 1956. Is Consciousness a Brain Process? *British Journal of Psychology* 47: 44–50.

Polger, T. 2002. Putnam's Intuition. *Philosophical Studies* 109 (2): 143–70.

Polger, T. 2004a. *Natural Minds*. Cambridge, MA: MIT Press.

Polger, T. 2004b. Neural Machinery and Realization. *Philosophy of Science* 71 (5): 997–1006.

Polger, T. 2007a. Realization and the Metaphysics of Mind. *Australasian Journal of Philosophy* 85 (2): 233–59.

Polger, T. 2007b. Some Metaphysical Anxieties of Reductionism. In M. Schouten and H. Looren de Jong (eds), *The Matter of the Mind: Philosophical Essays on Psychology, Neuroscience and Reduction*. Oxford: Blackwell.

Polger, T. 2008. Computational Functionalism. In P. Calvo and J. Symons (eds), *The Routledge Companion to the Philosophy of Psychology*. London: Routledge.

Polger, T. 2009a. Evaluating the Evidence for Multiple Realization. *Synthese* 167 (3): 457–72.

Polger, T. 2009b. Two Confusions Concerning Multiple Realizability. *Philosophy of Science* 75 (5): 537–47.

Polger, T. 2010. Mechanisms and Explanatory Realization Relations. *Synthese* 177 (2): 193–212.

Polger, T. 2011. Are Sensations Still Brain Processes? *Philosophical Psychology* 24 (1): 1–21.

Polger, T. 2012. Functionalism as a Philosophical Theory of the Cognitive Sciences. *WIRE Cognitive Science* 3: 337–48.

Polger, T. 2015. Realization and Multiple Realization, Chicken and Egg. *European Journal of Philosophy* 23 (4): 862–77.

Polger, T. and L. Shapiro. 2008. Understanding the Dimensions of Realization. *Journal of Philosophy*, CV (4): 213–22.

Polger, T. and K. Sufka. 2006. Closing the Gap on Pain: Mechanism, Theory, and Fit. In M. Aydede (ed.), *New Essays on the Nature of Pain and the Methodology of Its Study*. Cambridge, MA: MIT Press.

Potochnik, A. 2010. Levels of Explanation Reconceived. *Philosophy of Science* 77 (1): 59–72.

Potochnik, A. 2012. Modeling Social and Evolutionary Games. *Studies in History and Philosophy of Biological and Biomedical Sciences* 43 (1): 202–8.

Potochnik, A. Forthcoming. *Idealization and the Aims of Science*. Chicago, IL: University of Chicago Press.

Price, C. J. and K. J. Friston. 2002. Degeneracy and Cognitive Anatomy. *Trends in Cognitive Sciences* 6 (10): 416–21.

Price, C. J., and K. J. Friston. 2005. Functional Ontologies for Cognition: The Systematic Definition of Structure and Function. *Cognitive Neuropsychology* 22 (3): 262–75.

Purves, D. and R. Lotto. 2003. *Why We See What We Do: An Empirical Theory of Vision*. Sunderland, MA: Sinauer Associates.

Purves, D., G. Augustine, D. Fitzpatrick et al. (eds). 2001. *Neuroscience*, 2nd ed. Sunderland, MA: Sinauer Associates.

Putnam, H. 1967. Psychological Predicates. In W. H. Capitan and D. D. Merrill (eds), *Art, Mind, and Religion*. Pittsburgh, PN: University of Pittsburgh Press. Reprinted as "The Nature of Mental States" in Putnam 1975.

Putnam, H. 1973. *Philosophy and Our Mental Life*. Berkeley, CA: Symposium. Reprinted in Putnam 1975.

Putnam, H. 1975. *Mind, Language, and Reality: Philosophical Papers*, Volume 2. Cambridge: Cambridge University Press.

Putnam, H. 1988. *Representation and Reality*. Cambridge, MA: MIT Press.

Putnam, H. 1999. *The Threefold Cord: Mind, Body, and World*. New York: Columbia University Press.

Pylyshyn, Z. 1984. *Computation and Cognition: Toward a Foundation for Cognitive Science*. Cambridge, MA: MIT Press.

Rey, G. 1997. *Contemporary Philosophy of Mind: A Contentiously Classical Approach*. Oxford: Blackwell.

Richardson, R. 1979. Functionalism and Reductionism. *Philosophy of Science* 46: 533–58.

Richardson, R. 1982. How Not to Reduce a Functional Psychology. *Philosophy of Science* 49 (1): 125–37.

Richardson, R. 2008. Autonomy and Multiple Realization. *Philosophy of Science* 75 (5): 526–36.

Richardson, R. 2009. Multiple Realization and Methodological Pluralism. *Synthese* 167 (3): 473–92.

Roe, A., S. Pallas, J. Hahm, and M. Sur. 1990. A Map of Visual Space Induced in Primary Auditory Cortex. *Science* 250: 818–20.

Roediger, H. L. III, E. J. Marsh, and S. C. Lee. 2002. Varieties of Memory. In D.L. Medin and H. Pashler (eds), *Stevens' Handbook of Experimental Psychology*, Volume 2, 3rd ed. New York: John Wiley and Son: 1–41.

Rosenberg, A. 2001. On Multiple Realization and the Special Sciences. *Journal of Philosophy* 98 (7): 365–73.

Rumelhart, D., J. McClelland, and the PDP Research Group. 1986. *Parallel Distributed Processing: Explorations in the Microstructure of Cognition*, Volume I. Cambridge, MA: MIT Press.

Sachse, C. and M. Esfeld. 2007. Theory Reduction by Means of Functional Sub-Types. *International Studies in the Philosophy of Science* 21 (1): 1–17.

Samuels, R. 1998. Evolutionary Psychology and the Massive Modularity Hypothesis. *British Journal of Philosophy of Science* 49 (4): 575–602.

Schacter, D. 1996. *Searching for Memory: The Brain, the Mind, and the Past*. New York: Basic Books.

Searle, J. 1980. Minds, Brains, and Programs. *Behavioral and Brain Sciences* 3 (3): 417–24.

Searle, J. 1992. *Rediscovery of Mind*. Cambridge, MA: MIT Press.

Seidenberg, M. and J. McClelland. 1989. A Distributed, Developmental Model of Word Recognition and Naming. *Psychological Review* 96: 523–68.

Seidenberg, M. and J. McClelland. 1990. More Words but Still No Lexicon: Reply to Besner et al. *Psychological Review* 97: 447–52.

Sellars, W. 1963. *Empiricism and the Philosophy of Mind*. London: Routledge and Kegan Paul.

Shagrir, O. 1998. Multiple Realization, Computation and the Taxonomy of Psychological States. *Synthese* 114 (3): 445–61.

Shagrir, O. 2006. Why We View the Brain as a Computer. *Synthese* 153 (3): 393–416.

Shagrir, O. 2012. Computation, Implementation, Cognition. *Minds and Machines* 22: 137–48.

Shapiro, L. 2000. Multiple Realizations. *Journal of Philosophy* 97 (12): 635–54.

Shapiro, L. 2004. *The Mind Incarnate*. Cambridge, MA: MIT Press.

Shapiro, L. 2008. How to Test for Multiple Realization. *Philosophy of Science* 75 (5): 514–25.

Shapiro, L. 2010. Lessons from Causal Exclusion. *Philosophy and Phenomenological Research* 81: 594–604.

Shapiro, L. 2011a. Mental Manipulations and the Problem of Causal Exclusion. *Australasian Journal of Philosophy* 90 (3): 507–24.

Shapiro, L. 2011b. *Embodied Cognition*. New York: Routledge.

Shapiro, L. Forthcoming. Mechanism or Bust? Explanation in Psychology. *British Journal for the Philosophy of Science*.

Shapiro, L. and T. Polger. 2012. Identity, Variability, and Multiple Realizability. In S. Gozzano and C. Hill (eds), *The Mental and the Physical: New Perspectives on Type Identity*. Cambridge: Cambridge University Press.

Shapiro, L. and E. Sober. 2007. Epiphenomenalism: The Dos and Don'ts. In G. Wolters and P. Machamer (eds), *Thinking about Causes: From Greek Philosophy to Modern Physics*. Pittsburgh, PN: University of Pittsburgh Press: 235–64.

Sharma, J., A. Angelucci, and M. Sur. 2000. Induction of Visual Orientation Modules in Auditory Cortex. *Nature* 404: 841–7.

Sharpe, L. T., A. Stockman, H. Jägle, and J. Nathans. 1999. Opsin Genes, Cone Photopigments, Color Vision and Colorblindness. In K. Gegenfurtner and L. T. Sharpe (eds), *Color Vision: From Genes to Perception*. Cambridge: Cambridge University Press: 3–51.

Shichida, Y. and T. Matsuyama. 2009. Evolution of Opsins and Phototransduction. *Philosophical Transactions of the Royal Society of London B: Biological Sciences* 364 (1531): 2881–95.

Shoemaker, S. 1975. Functionalism and Qualia. *Philosophical Studies* 27: 291–315.

Shoemaker, S. 1980. Causality and Properties. In P. van Inwagen (ed.) *Time and Cause*. Dordrecht: D. Reidel Publishing. Reprinted in Shoemaker 1984.

Shoemaker, S. 1981. Some Varieties of Functionalism. *Philosophical Topics* 12 (1): 83–118. Reprinted in Shoemaker 1984.

Shoemaker, S. 1984. *Identity, Cause, and Mind*. New York: Cambridge University Press.

Sider, T. 2003. What's so Bad about Overdetermination? *Philosophy and Phenomenological Research* 67 (3): 719–26.

Simon, H. 1962. The Architecture of Complexity. *Proceedings of the American Philosophical Society* 106 (6): 467–82.

Simon, H. 1996. The Sciences of the Artificial, 3rd ed. Cambridge, MA: MIT Press.

Skinner, B. F. 1953. *Science and Human Behavior*. New York: Macmillan.

Skipper, R. and R. Millstein. 2005. Thinking about Evolutionary Mechanisms: Natural Selection. *Studies in History and Philosophy of Biological and Biomedical Sciences* 36 (2): 327–47.

Smart J. J. C. 1959. Sensations and Brain Processes. *Philosophical Review* LXVIII: 141–56.

Smart, J. J. C. 1961. Further Remarks on Sensations and Brain Processes. *Philosophical Review* 70 (3): 406–7.

Smart, J. J. C. 2007. The Identity Theory of Mind. *Stanford Encyclopedia of Philosophy*, <http://plato.stanford.edu/entries/mind-identity/>.

Smith, E. and G. Lewin. 2009. Nociceptors: A Phylogenetic View. *Journal of Comparative Physiology A* 195: 1089–106.

Smith. M. 1994. *The Moral Problem*. Oxford: Wiley-Blackwell.

Smolensky, P. 1988. On the Proper Treatment of Connectionism. *Behavioral and Brain Sciences* 11: 1–74.

Sober, E. 1990/1985. Panglossian Functionalism and the Philosophy of Mind. *Synthese* 64: 165–93. In Lycan 1990.

Sober, E. 1999. The Multiple Realizability Argument against Reductionism. *Philosophy of Science* 66: 542–64.

Sober, E. 2008. *Evidence and Evolution: The Logic behind the Science*. Cambridge: Cambridge University Press.

Soom, P. 2012. Mechanisms, Determination and the Metaphysics of Neuroscience. *Studies in History and Philosophy of Science Part C* 43 (3): 655–64.

Soom, P., C. Sachse, and M. Esfeld. 2010. Functional Sub-Types. *Journal of Consciousness Studies* 17 (1–2): 1–2.

Stillings, N. A., S. E. Weisler, C. H. Chase, M. H. Feinstein, J. L. Garfield, and E. L. Rissland. 1987. *Cognitive Science: An Introduction*. Cambridge, MA: MIT Press.

Sullivan, J. 2008. Memory Consolidation, Multiple Realization and Modest Reductions. *Philosophy of Science* 75 (5): 501–13.

Sullivan, J. 2009. The Multiplicity of Experimental Protocols: A Challenge to Reductionist and Non-Reductionist Models of the Unity of Neuroscience. *Synthese* 167 (3): 511–39.

Sullivan, J. 2010. Reconsidering "Spatial Memory" and the Morris Water Maze. *Synthese* 177 (2): 261–83.

Sullivan, J. 2016. Neuroscientific Kinds through the Lens of Scientific Practice. In C. Kendig (ed.), *Natural Kinds and Classification in Scientific Practice*. London: Pickering and Chatto.

Suppe, F. 1989. *The Semantic Conception of Theories and Scientific Realism*. Urbana, IL: University of Illinois Press.

Surridge, A., D. Osorio, and N. Mundy. 2003. Evolution and Selection of Trichromatic Vision in Primates. *Trends in Ecology and Evolution* 18 (4): 198–205.

Thaler, S., S. Arnott, and M. Goodale. 2011. Neural Correlates of Natural Human Echolocation in Early and Late Blind Echolocation Experts. *PLoS One* 25: e20162.

Theurer, K. 2013. Compositional Explanatory Relations and Mechanistic Reduction. *Minds and Machines* 23 (3): 287–307.

Thomasson, A. 1999. *Fiction and Metaphysics*. Cambridge: Cambridge University Press.

Thompson, R. and J. Kim. 1996. Memory Systems in the Brain and Localization of a Memory. *Proceedings of the National Academy of Sciences of the United States of America* 93 (24): 13438–44.

Tomberlin, J. (ed.). 1997. *Philosophical Perspectives 11: Mind, Causation, and World*. Boston, MA: Blackwell.

Tononi, G., O. Sporns, and G. M. Edelman. 1999. Measures of Degeneracy and Redundancy in Biological Networks. *Proceedings of the National Academy of Sciences* 96: 3257–62.

Towl, B. 2012. Laws and Constrained Kinds: A Lesson from Motor Neuroscience. *Synthese* 189 (3): 433–50.

Tulving, E. 1972. Episodic and Semantic Memory. In E. Tulving and W. Donaldson (eds), *Organization of Memory*. New York: Academic Press: 381–402.

Turvey, M. T. 1977. Preliminaries to a Theory of Action with Reference to Vision. In R. Shaw and J. Bransford (eds), *Perceiving, Acting and Knowing*. Hillsdale, NJ: Erlbaum: 211–65.

Ugural, A. and S. Fenster. 2003. *Advanced Strength and Applied Elasticity*, 4th edition. Upper Saddle River, NJ: Pearson Education.

van Fraassen, B. 1980. *The Scientific Image*. Oxford: Oxford University Press.

Varela, F., E. Thompson, and E. Rosch. 1991. *The Embodied Mind: Cognitive Science and Human Experience*. Cambridge, MA: MIT Press.

Vogel, S. 2013. *Comparative Biomechanics: Life's Physical World*. Princeton, NJ: Princeton University Press.

Von Eckardt, B. 1993. *What Is Cognitive Science?* Cambridge, MA: MIT Press.

von Melchner, L., S. Pallas, and M. Sur. 2000. Visual Behaviour Mediated by Retinal Projections Directed to the Auditory Pathway. *Nature* 404: 871–6.

Walter, S. 2006. Multiple Realizability and Reduction: A Defense of the Disjunctive Move. *Metaphysica* 7 (1): 43–65.

Walter, S. 2008. The Supervenience Argument, Overdetermination, and Causal Drainage: Assessing Kim's Master Argument. *Philosophical Psychology* 21 (5): 673–96.

Walter, S. and M. Eronen. 2011. Reduction, Multiple Realizability, and Levels of Reality. In S. French and J. Saatsi (eds), *Continuum Companion to the Philosophy of Science*. London: Continuum: 138–56.

Watson, J. 1913. Psychology as the Behaviorist Views It. *Psychological Review* 20: 158–77.

Weisberg, M. 2007. Three Kinds of Idealization. *Journal of Philosophy* 104 (12): 639–59.

Weisberg, M. 2013. *Simluation and Similarity*. New York: Oxford University Press.

Weiskopf, D. 2011. The Functional Unity of Special Science Kinds. *British Journal for the Philosophy of Science* 62 (2): 233–58.

Weiskopf, D. Forthcoming. The Explanatory Autonomy of Cognitive Models. In D. M. Kaplan (ed.), *Integrating Psychology and Neuroscience: Prospects and Problems*. Oxford: Oxford University Press.

Weslake, B. 2010. Explanatory Depth. *Philosophy of Science* 77 (2): 273–94.

Wilson, J. 2009. Determination, Realization and Mental Causation. *Philosophical Studies* 145: 149–69.

Wilson, J. 2010. Non-Reductive Physicalism and Degrees of Freedom. *British Journal for the Philosophy of Science* 61 (2): 279–311.

Wimsatt, W. 1976. Reductionism, Levels of Organization, and the Mind-Body Problem. In G. Globus, I. Savodnik, and G. Maxwell (eds), *Consciousness and the Brain*. New York: Plenum Press: 199–267.

Wimsatt, W. 2006. Reductionism and Its Heuristics: Making Methodological Reductionism Honest. *Synthese* 151 (3): 445–75.

Wimsatt, W. 2007. *Re-Engineering Philosophy for Limited Beings: Piecewise Approximations to Reality*. Cambridge, MA: Harvard University Press.

Wimsatt, W. 2013. Evolution and the Stability of Functional Architectures. In P. Huneman (ed.), *Functions: Selection and Mechanisms*. Dordrecht: Springer: 19–41.

Witmer, G. 2003. Functionalism and Causal Exclusion. *Pacific Philosophical Quarterly* 84 (2): 198–214.

Wong, M. and A. Bhattacharjee. 2011. How Does the Visual Cortex of the Blind Acquire Auditory Responsiveness? *Frontiers in Neuroanatomy* 5: 52.

Woodward, J. 2000. Explanation and Invariance in the Special Sciences. *British Journal for the Philosophy of Science* 51: 197–254.

Woodward, J. 2003. *Making Things Happen: A Theory of Causal Explanation*. Oxford: Oxford University Press.

Woodward, J. 2008. Mental Causation and Neural Mechanisms. In J. Hohwy and J. Kallestrup (eds), *Being Reduced: New Essays on Reduction, Explanation, and Causation*. Oxford: Oxford University Press: 218–62.

Yablo, S. 1992. Mental Causation. *Philosophical Review* 101: 245–80.

Index

Obama, Barack 213–14
occupant, *see* realization
octopus 7, 14, 31, 34, 38, 40–5, 57, 60–6,
79, 83–5, 99, 102, 111–17, 137–9,
179, 190
only game in town argument 150–1, 158
ontology 13, 19–22, 27, 29, 135, 151, 154–7,
161–2, 166–7, 175, 178, 182, 184–5,
188–9, 194, 200, 203–4, 222, 226
Oppenheim, Paul 199, 213, 217–18,
235, 240
opsin 99, 104, 106–11, 179–80, 225, *see
also* eye, vision, trichromacy
optic 42, 44–5, 62, 66, 73–4, 85, 160,
168, 218, *see also* eye
organ 42, 44–5, 62–3, 73–4, 86–9,
118–19, 137, 179, 188
Orlandi, Nico 153, 160, 240
oscillator 70, 142
otter 83, 138–9
output 19, 24, 46–7, 96, 108, 119, 121,
133, 153–5, 165
overdetermination 193, 197, *see also*
Exclusion Principle
oxygen 88

pain 6–8, 14, 17, 34, 38, 41–2, 55, 77, 83,
94, 102, 111–14, 123, 136, 144, 168,
189–90, 192, 223, *see also*
nociception
painting 20
pallium 115–16, *see also* dorsal ventral
ridge
Palmer, David 5, 240
pancomputationalism 161, *see also*
computation
paperweight 68
Papineau, David 196, 240
Parker, Sarah xi
parrot 138
part 5, 20, 28, 30, 39, 42, 61, 79, 90, 97–8,
106–8, 111–13, 118, 132, 142–4,
165–6, 188, 208
particle 8, 26, 29, 72, 78
Pascalis, Olivier 48–9, 240
Pauen, Michael 27, 239–40
peacock 39–40
Pearl, Judea 210, 240
peg 8–9, 14, 201–2, 206
Pelger, Susanne 73, 239
peppermint 65
perception 4, 48–9, 86, 93–4, 97–9,
109–10, 143–4, 160, 183, 192

periodic table 68, 70
permissibility 100, 103, 131, 188
Pettit, Philip 12–13, 236
phylogeny 120, 128, 181
physicalism ix, 3–5, 11, 14, 17–19, 21,
37, 41–2, 45, 58, 72, 78, 84, 96, 107,
124, 130, 132, 148, 196–7, 200, 209,
221, 228
physics 4, 24, 37, 55, 78, 195, 203, 227
physiology 87–9, 114–15, 117, 119, 128,
132, 136–7, 146, 148, 153–4
Piatigorsky, Joram 73, 240
Piccinini, Gualtiero xi, 149, 157, 160–6,
227, 240
plasticity, neural and synaptic 44, 83, 86,
88–99, 111, 114, 123, 129–33,
135–6, 145, 149, 167, 191, 225
pluralism 13, 206–10, 216–19
Polger, Thomas 11, 21, 23, 29–30, 33–4,
39, 50–1, 67, 77, 79, 90, 112, 153,
170, 204, 222–5, 236, 240–1, 243
polymorphism 88, 107
posit 5, 30, 46, 107, 154, 182, 199
possibility 14, 18, 23–4, 40, 42, 48, 50,
52, 55–9, 61, 77, 85, 121, 128,
134–7, 144, 146–9, 168, 200,
223, 226
Postulate 184–6, *see also* strategies for
dealing with Abstraction and
Idealization
potassium 39
Potochnik, Angela 166, 170, 183, 186,
206, 213, 217, 227, 241
powers 15, 28, 70, 196, 198, 206
pragmatism 205
Preston, Dawn xi
Probability 7, 23, 32, 43, 125–6, *see also*
likelihood
properties x, 10, 20, 22–5, 27–31, 33, 53,
56–7, 65, 71, 73, 87–8, 102, 107,
114–15, 120–1, 132–6, 143, 152,
156, 163–6, 181, 196, 202–3, 210,
213, 217, 222
protein 78, 104, 106–7
protons 68
psychology ix, xi, 3–27, 30–42, 44–62,
77–104, 110–14, 121–8, 161, 167–9,
175–86, 211, 215–18, 221–7
psychophysical 128, 146, 195
pulleys 6, 8
Purves, Dale 106, 160, 241
Putnam, Hilary 6–9, 11–16, 21–3, 32–4,
37, 40–5, 51, 53–4, 56–7, 78, 83–5,